Fourier Series and Integral Transforms

Fourier Series and Integral Transforms

Allan Pinkus & Samy Zafrany
Technion, Israel Institute of Technology

PUBLISHED BY THE PRESS SYNDICATE OF THE UNIVERSITY OF CAMBRIDGE
The Pitt Building, Trumpington Street, Cambridge CB2 1RP, United Kingdom

CAMBRIDGE UNIVERSITY PRESS
The Edinburgh Building, Cambridge, CB2 2RU, United Kingdom
40 West 20th Street, New York, NY 10011-4211, USA
10 Stamford Road, Oakleigh, Melbourne 3166, Australia

© Cambridge University Press 1997

This book is in copyright. Subject to statutory exception
and to the provisions of relevant collective licensing agreements,
no reproduction of any part may take place without
the written permission of Cambridge University Press.

First published 1997

Printed in the United Kingdom at the University Press, Cambridge

A catalogue record for this book is available from the British Library

ISBN 0 521 59209 7 hardback
ISBN 0 521 59771 4 paperback

Contents

Preface vii

0 Notation and Terminology 1
    1   Basic Concepts in Set Theory 1
    2   Calculus Notation 2
    3   Useful Trigonometric Formulae 4

1 Background: Inner Product Spaces 5
    0   Introduction 5
    1   Linear and Inner Product Spaces 5
    2   The Norm 10
    3   Orthogonal and Orthonormal Systems 15
    4   Orthogonal Projections and Approximation
         in the Mean 19
    5   Infinite Orthonormal Systems 24
         Review Exercises 30

2 Fourier Series 32
    0   Introduction 32
    1   Definitions 32
    2   Evenness, Oddness, and Additional Examples 40
    3   Complex Fourier Series 42
    4   Pointwise Convergence and Dirichlet's Theorem 46
    5   Uniform Convergence 56
    6   Parseval's Identity 63
    7   The Gibbs Phenomenon 68
    8   Sine and Cosine Series 72
    9   Differentiation and Integration of Fourier Series 76
   10   Fourier Series on Other Intervals 81
   11   Applications to Partial Differential Equations 85
         Review Exercises 89

3 The Fourier Transform 93
    0   Introduction 93
    1   Definitions and Basic Properties 93

	2	Examples	98
	3	Properties and Formulae	102
	4	The Inverse Fourier Transform and Plancherel's Identity	108
	5	Convolution	116
	6	Applications of the Residue Theorem	119
	7	Applications to Partial Differential Equations	125
	8	Applications to Signal Processing	130
		Review Exercises	137
4		**The Laplace Transform**	**140**
	0	Introduction	140
	1	Definition and Examples	140
	2	More Formulae and Examples	143
	3	Applications to Ordinary Differential Equations	149
	4	The Heaviside and Dirac-Delta Functions	155
	5	Convolution	162
	6	More Examples and Applications	168
	7	The Inverse Transform Formula	173
	8	Applications of the Inverse Transform	175
		Review Exercises	180

Appendix A: The Residue Theorem and Related Results .. 182

Appendix B: Leibniz's Rule and Fubini's Theorem 186

Index 188

Preface

The aim of this book is to provide the reader with a basic understanding of Fourier series, Fourier transforms, and Laplace transforms. Fourier series (and power series) are important examples of useful series of functions. Applications of Fourier series may be found in many diverse theoretical and applied areas. The same holds for integral transforms. The Fourier and Laplace transforms are the best known of these transforms and are prototypes of the general integral transforms.

Fourier series and integral transforms are theoretically based on a natural amalgamation of concepts from both linear algebra and integral and differential calculus. In other words, they are a mix of algebra and analysis. We assume that the reader is well versed in the basics of these two areas. Nevertheless, in Chapter 1 is found a somewhat concise review of some of the relevant concepts and facts from linear algebra.

The best, most efficient, and perhaps only way to learn mathematics is to study and review the material and to solve exercises. At the end of almost every section of this book may be found a collection of exercises. A set of review exercises is to be found at the end of each chapter. To truly and properly understand the subject matter, it is essential to solve exercises.

This book was originally prepared, in Hebrew, for the course "Fourier Series and Integral Transformations" given by the Department of Mathematics at the Technion – Israel Institute of Technology, Haifa, Israel. This course is intended for students of the Department of Electrical Engineering, but also includes students of the Physics and Computer Science Departments. It is a one semester course (14 weeks), given during the third semester of studies, with two hours of lectures and a one hour problem session each week. The material presented in this book is an expanded, and hopefully polished, version of the notes for this course.

We would like to thank all those who helped in the preparation of this book, and who found and corrected numerous misprints. Naturally, all remaining errors are totally our own.

We very much hope that the material of this book is presented in a manner which enables you, the reader, to digest the subject matter efficiently and thoroughly. We wish you every success in this endeavour.

Allan Pinkus,   Samy Zafrany

Chapter 0

Notation and Terminology

1. Basic Concepts in Set Theory

We assume that the reader is conversant with the simple concepts of set theory. This being so, we present only a few of the basic definitions and notations to set the tone for what follows.

General set theory deals with all sorts of sets. But mathematical analysis, as studied in undergraduate courses, generally deals with sets of numbers as well as sets of real- or complex-valued functions. A set is a collection of *elements*. We usually denote the elements by lower case letters: $a, b, c, \ldots$. If a belongs to the set A, then we say that a is an *element* of A and write $a \in A$. If b is *not* an element of A, we write $b \notin A$.

The set A composed of the elements $-6, 3+i, 5i-2$, and 7 we denote by

$$A = \{-6,\ 3+i,\ 5i-2,\ 7\}.$$

In this same way we can describe a set of functions

$$B = \{\sin x,\ \sin 2x,\ \sin 3x,\ \ldots,\ \sin 100x\}$$

as the set which contains exactly 100 elements, each of which is a function. In a similar fashion we also denote an infinite set of functions. As an example we have

$$\{1,\ 1+2x,\ 1+2x+3x^2,\ 1+2x+3x^2+4x^3,\ \ldots\}.$$

In general, we describe a set by writing

(0.1) $$A = \{x \mid \phi(x)\}$$

where the variable x denotes the elements of the known set (be it numbers or functions) while the expression $\phi(x)$ represents a property which must be satisfied by the element x. The $\phi(x)$ will often be an equation or an order relation with respect to x. The meaning of definition (0.1) is that A is the set of all elements x which satisfy property ϕ. For example, consider the set

$$A = \left\{ x \mid x^4 - 1 = 0 \right\}.$$

Assuming that x is a complex number, we can rewrite A as

$$A = \{1,\ -1,\ i,\ -i\}.$$

It is sometimes more convenient to express the property in words. Consider, for example,

$$A = \{\,x \mid x \text{ is an odd integer}\,\},$$

$$\mathcal{P}_3 = \{\,p \mid p \text{ is a polynomial of degree at most 3}\,\}.$$

2. Calculus Notation

The positive integers, i.e., the numbers 1, 2, 3, ..., are called the *natural numbers*. The set of natural numbers we denote by $\mathbb{N}$. Thus

$$\mathbb{N} = \{1,\ 2,\ 3,\ 4,\ \ldots\}.$$

The set of all integers we denote by $\mathbb{Z}$. Hence

$$\mathbb{Z} = \{\ldots,\ -3,\ -2,\ -1,\ 0,\ 1,\ 2,\ 3,\ \ldots\}.$$

The set of all non-negative integers (including 0) we denote by $\mathbb{Z}_+$. Thus

$$\mathbb{Z}_+ = \{0,\ 1,\ 2,\ 3,\ \ldots\}.$$

The sole difference between $\mathbb{Z}_+$ and $\mathbb{N}$ is that the former contains 0, while the latter does not. Every number of the form $\frac{m}{n}$, where m and n are integers, and $n \neq 0$, is called a *fraction* or a *rational number*. The set of all rational numbers is denoted by $\mathbb{Q}$. Thus

$$\mathbb{Q} = \left\{ \frac{m}{n} \;\middle|\; m \text{ and } n \text{ are integers and } n \neq 0 \right\}.$$

The set of all real numbers we denote by $\mathbb{R}$. Hence

$$\mathbb{R} = \{\,x \mid x \text{ is a real number}\,\}.$$

The set of non-negative real numbers is $\mathbb{R}_+$. That is

$$\mathbb{R}_+ = \{\,x \mid x \in \mathbb{R}, x \geq 0\,\}.$$

Chapter 0: Notation and Terminology

Finally $\mathbb{C}$ denotes the set of complex numbers. Thus

$$\mathbb{C} = \{x + iy \mid x, y \in \mathbb{R}\}.$$

When we write a complex number $z \in \mathbb{C}$ in the form $z = x + iy$, $x, y \in \mathbb{R}$, x is called the *real part* of z and y the *imaginary part* of z. We use the notation $x = \text{Re}(z)$ and $y = \text{Im}(z)$.

In this book we use the following standard notation for real intervals: For each two real numbers a and b we define:

$$\text{Open interval:} \quad (a, b) = \{x \mid a < x < b\}$$
$$\text{Closed interval:} \quad [a, b] = \{x \mid a \leq x \leq b\}$$
$$\text{Half-closed half-open interval:} \quad [a, b) = \{x \mid a \leq x < b\}$$
$$\text{Half-open half-closed interval:} \quad (a, b] = \{x \mid a < x \leq b\}$$

These sets are all the *finite* intervals of $\mathbb{R}$. There are also *infinite* intervals. For each real a we define four different intervals:

$$\text{Left-open interval:} \quad (a, \infty) = \{x \mid a < x\}$$
$$\text{Left-closed interval:} \quad [a, \infty) = \{x \mid a \leq x\}$$
$$\text{Right-open interval:} \quad (-\infty, a) = \{x \mid x < a\}$$
$$\text{Right-closed interval:} \quad (-\infty, a] = \{x \mid x \leq a\}$$

The set of all the real numbers can also be denoted as

$$\mathbb{R} = (-\infty, \infty) = \{x \mid -\infty < x < \infty\}.$$

The character ∞ (and $-\infty$) is the symbolic expression for infinity (and minus infinity). It is not a number.

A function $f : [a, b] \to \mathbb{C}$ is said to be *piecewise continuous* if it has at most a finite number of points of discontinuity and if, in addition, at each point of discontinuity both one-sided limits exist (and are finite). We wish to emphasize that we do not require that the function be defined at its points of discontinuity. Every such function f may be written in the form

$$f = u + iv$$

where u and v are real-valued piecewise continuous functions on the interval $[a, b]$. The function u is called the *real part* of f and we write $u = \text{Re}(f)$. The function v is called the *imaginary part* of f and we write $v = \text{Im}(f)$. By $C[a,b]$ we denote the set of all continuous functions $f : [a,b] \to \mathbb{C}$. By $E[a,b]$ we denote the set of all piecewise continuous functions $f : [a,b] \to \mathbb{C}$.

3. Useful Trigonometric Formulae

$$\sin(\pi - \alpha) = \sin \alpha \qquad \sin n\pi = 0, \quad n \in \mathbb{Z}$$

$$\cos(\pi - \alpha) = -\cos \alpha \qquad \cos n\pi = (-1)^n, \quad n \in \mathbb{Z}$$

$$\sin \alpha = \cos(\tfrac{\pi}{2} - \alpha) \qquad \tan \alpha = \cot(\tfrac{\pi}{2} - \alpha)$$

$$\tan \alpha = \frac{\sin \alpha}{\cos \alpha} \qquad \cot \alpha = \frac{\cos \alpha}{\sin \alpha}$$

$$\sin^2 \alpha + \cos^2 \alpha = 1$$

$$\sin(\alpha \pm \beta) = \sin \alpha \cos \beta \pm \cos \alpha \sin \beta$$

$$\cos(\alpha \pm \beta) = \cos \alpha \cos \beta \mp \sin \alpha \sin \beta$$

$$\sin 2\alpha = 2 \sin \alpha \cos \alpha$$

$$\cos 2\alpha = \cos^2 \alpha - \sin^2 \alpha$$

$$\sin \alpha \sin \beta = \tfrac{1}{2}[\cos(\alpha - \beta) - \cos(\alpha + \beta)]$$

$$\sin \alpha \cos \beta = \tfrac{1}{2}[\sin(\alpha + \beta) + \sin(\alpha - \beta)]$$

$$\cos \alpha \cos \beta = \tfrac{1}{2}[\cos(\alpha + \beta) + \cos(\alpha - \beta)]$$

$\sin^2 \alpha = \tfrac{1}{2} - \tfrac{1}{2} \cos 2\alpha \qquad\qquad \cos^2 \alpha = \tfrac{1}{2} + \tfrac{1}{2} \cos 2\alpha$

$\sin^3 \alpha = \tfrac{1}{4}(3 \sin \alpha - \sin 3\alpha) \qquad \cos^3 \alpha = \tfrac{1}{4}(3 \cos \alpha + \cos 3\alpha)$

$\sin^4 \alpha = \tfrac{1}{8}(3 - 4 \cos 2\alpha + \cos 4\alpha) \qquad \cos^4 \alpha = \tfrac{1}{8}(3 + 4 \cos 2\alpha + \cos 4\alpha)$

Chapter 1
Background: Inner Product Spaces

0. Introduction

The main topics to be studied in this chapter are orthogonal and orthonormal systems in a vector space with inner product, as well as various related concepts. These topics are sometimes, but not always, discussed in a basic course in linear algebra. Of central importance is the subject of infinite orthonormal systems which we present at the end of this chapter. These results will be applied in the next chapter on Fourier series. The first four sections of this chapter are a condensed review of some concepts and basic ideas (with proofs) from linear algebra. We use these facts in developing the different topics of this book. The reader will hopefully find in these sections a helpful synopsis and review of his knowledge of the area.

1. Linear and Inner Product Spaces

The basic algebraic structure which we use is the *linear space* (often called *vector space*) over a *field of scalars*. Our "field of scalars" will always be either the real numbers $\mathbb{R}$ or the complex numbers $\mathbb{C}$. Elements of a linear space are called *vectors*. Formally, a non-empty set V is called a *linear space* over a field F if it satisfies the following conditions:

1. **Vector Addition:** There exists an operation, generally denoted by "+", such that for any two vectors $u, v \in V$, the "sum" $u + v$ is also a vector in V.
2. **Associativity:** For every $u, v, w \in V$, we have $(u + v) + w = u + (v + w)$.
3. **Zero Vector:** There exists a vector which we denote by $\vec{0}$ and call the "zero vector" satisfying $\vec{0} + v = v$, for all $v \in V$.

4. **Inverse vector:** For every vector $v \in V$ there exists a vector, which we denote by $-v$, and call "minus v", such that $v + (-v) = \vec{0}$.
5. **Commutativity:** For every $u, v \in V$ we have $u + v = v + u$.
6. **Multiplication by a Scalar:** Multiplication by scalars is permissible. That is, for each $v \in V$ and scalar $a \in F$, there is defined $av \in V$.
7. For every $a \in F$ and $u, v \in V$, $a(u + v) = au + av$.
8. For each $a, b \in F$ and $u \in V$, $(a + b)u = au + bu$ and $a(bu) = (ab)u$.
9. For the unit scalar 1 of F and every $v \in V$, $1 \cdot v = v$.

If V is a linear space over the field of reals $\mathbb{R}$, then we say that V is a *real linear space*. If V is a linear space over the field of complex numbers $\mathbb{C}$, then we call V a *complex linear space*. A subset W of V ($W \subseteq V$) is said to be a *linear subspace* of the space V if all the above conditions hold for W over the same field of scalars as for V. Of course, the operations of vector addition and multiplication by a scalar must be the same in W as in V. A well-known criterion for checking if W is a subspace is the following: $W \neq \emptyset$ and for every $u, v \in W$ and $a, b \in F$ we have $au + bv \in W$. In other words, W is a linear subspace of V if and only if W is a non-empty subset of V, which is closed under the operation of vector addition and multiplication by scalars. We assume in what follows, unless stated otherwise, that all our linear subspaces are complex. We now quickly review a number of important concepts related to linear spaces.

Definition 1.1: *Let V be a linear space and $v_1, \ldots, v_n \in V$. The vector u is said to be a* linear combination *of the vectors $v_1, \ldots, v_n$ if there exist scalars $a_1, \ldots, a_n \in F$ such that*

$$u = a_1 v_1 + a_2 v_2 + \cdots + a_n v_n.$$

The collection of all vectors u which are linear combinations of $v_1, \ldots, v_n$ is called the span *of $v_1, \ldots, v_n$ and will be denoted by* $\mathrm{span}\{v_1, \ldots, v_n\}$.

We also say that $v_1, \ldots, v_n$ *span* W. Note that $W = \mathrm{span}\{v_1, \ldots, v_n\}$ is a linear subspace of V.

Definition 1.2: *Let V be a linear space. The vectors $v_1, \ldots, v_n \in V$ are said to be* linearly independent *if the equation*

$$a_1 v_1 + a_2 v_2 + \cdots + a_n v_n = \vec{0}, \qquad a_1, a_2, \ldots, a_n \in F,$$

is satisfied only by the scalars $a_1 = \cdots = a_n = 0$. Otherwise we say that the vectors $v_1, \ldots, v_n$ are linearly dependent.

From the above two definitions it immediately follows that the vectors v_1, ..., v_n are linearly independent if and only if for each i, $1 \leq i \leq n$, the vector v_i is not a linear combination of the other vectors of the set.

Definition 1.3: *A finite set of vectors v_1, ..., v_n is said to be a* basis *for the linear space V if the set of vectors v_1, ..., v_n is linearly independent and $V = \text{span}\{v_1, \ldots, v_n\}$. The natural number n is called the dimension of V and we write $n = \dim(V)$.*

The reader will recall from his knowledge of linear algebra that every real or complex non-trivial linear space has an infinite number of different bases. However, any two bases have the same number of elements and thus the definition of dimension is in fact independent of any specific choice of basis (not necessarily finite in number). Every linear space has a dimension (which may be infinite).

Another important concept is that of an inner product. Note that the definition of a linear space does not include an operation of "multiplication" between vectors. The inner product could be considered as such an operation. However, linear spaces do not in general possess an inner product.

Definition 1.4: **(Inner Product)** *Let V be a real or complex linear space. An inner product is an operation between two elements of V which results in a scalar (and not a vector!). We denote it by $\langle u, v \rangle$, i.e., $\langle u, v \rangle \in \mathbb{C}$. It satisfies:*

1. *For each $v \in V$, $\langle v, v \rangle$ is a non-negative real number, i.e., $\langle v, v \rangle \geq 0$.*
2. *For each $v \in V$, $\langle v, v \rangle = 0$ if and only if $v = \vec{0}$.*
3. *For each $u, v, w \in V$ and $a, b \in F$, $\langle au + bv, w \rangle = a \langle u, w \rangle + b \langle v, w \rangle$.*
4. *For each $u, v \in V$ we have $\langle u, v \rangle = \overline{\langle v, u \rangle}$.*

A linear space with an inner product defined on it is called an inner product *space.*

The expression $\overline{\langle v, u \rangle}$ denotes the complex conjugate of the complex number $\langle v, u \rangle$. If our field of scalars is $\mathbb{R}$, then in place of Condition 4 we write $\langle u, v \rangle = \langle v, u \rangle$. There are numerous consequences of the above four conditions. Here are a few of them.

(a) For each $u, v, w \in V$ and $a, b \in \mathbb{C}$, $\langle u, av + bw \rangle = \overline{a} \langle u, v \rangle + \overline{b} \langle u, w \rangle$.

(b) For each $v \in V$ and $a \in \mathbb{C}$, $\langle av, av \rangle = |a|^2 \langle v, v \rangle$.

(c) For each $v \in V$, $\langle \vec{0}, v \rangle = 0$.

(d) In general, for each finite sequence of vectors $\{u_k\}_{k=1}^n$, scalars $\{a_k\}_{k=1}^n$, and every vector v,

$$\left\langle \sum_{k=1}^n a_k u_k, v \right\rangle = \sum_{k=1}^n a_k \langle u_k, v \rangle,$$

$$\left\langle v, \sum_{k=1}^n a_k u_k \right\rangle = \sum_{k=1}^n \overline{a_k} \langle v, u_k \rangle.$$

We now consider some typical examples of inner product spaces.

Example 1.1: The Euclidean space $V = \mathbb{R}^n$ with the usual operations of vector addition and multiplication by a scalar is a linear space over $\mathbb{R}$. Let $\mathbf{r} = (r_1, r_2, \ldots, r_n)$ be a vector of strictly positive numbers, i.e., $r_k > 0$, $1 \leq k \leq n$. We define an inner product $\langle \cdot, \cdot \rangle_{\mathbf{r}}$ on $\mathbb{R}^n$ in the following way: For each pair of vectors $\mathbf{x} = (x_1, x_2, \ldots, x_n)$ and $\mathbf{y} = (y_1, y_2, \ldots, y_n)$ in $\mathbb{R}^n$, set

$$\langle \mathbf{x}, \mathbf{y} \rangle_{\mathbf{r}} = \sum_{k=1}^n r_k x_k y_k.$$

The vector $\mathbf{r}$ is called the *weight vector* of the inner product. If $r_k = 1$, $1 \leq k \leq n$, then the inner product is denoted $\mathbf{x} \cdot \mathbf{y}$,

$$\mathbf{x} \cdot \mathbf{y} = x_1 y_1 + x_2 y_2 + \cdots + x_n y_n,$$

and is said to be the *standard inner product* (or *dot product*) on $\mathbb{R}^n$.

Example 1.2: Analogously to the previous example, $V = \mathbb{C}^n$ with the usual vector addition and multiplication by scalars is a linear space over $\mathbb{C}$. Let $\mathbf{r}$ be as in Example 1.1. For each pair $\mathbf{x}, \mathbf{y} \in \mathbb{C}^n$, we define

$$\langle \mathbf{x}, \mathbf{y} \rangle_{\mathbf{r}} = \sum_{k=1}^n r_k x_k \overline{y_k}.$$

It is not difficult to prove that this is an inner product on $\mathbb{C}^n$. As in the previous example, the *standard inner product* on $\mathbb{C}^n$ is

$$\mathbf{x} \cdot \mathbf{y} = x_1 \overline{y_1} + x_2 \overline{y_2} + \cdots + x_n \overline{y_n}.$$

Example 1.3: Let $V = C[a,b]$ be the space of continuous functions $f : [a,b] \to \mathbb{C}$ with the usual operations of sum of functions and multiplication by scalars. This is a linear space over $\mathbb{C}$. For each pair of functions $f, g \in C[a,b]$, we define

$$\langle f, g \rangle = \int_a^b f(x) \overline{g(x)} \, dx.$$

It is easy to verify that this is an inner product on $C[a,b]$.

Example 1.4: Set

$$\ell_2 = \left\{ \mathbf{x} \mid \mathbf{x} = (x_1, x_2, \ldots),\ x_n \in \mathbb{C},\ \sum_{n=1}^{\infty} |x_n|^2 < \infty \right\}.$$

That is, ℓ_2 is the space of infinite sequences of complex numbers $\{x_n\}_{n=1}^{\infty}$ for which the series $\sum_{n=1}^{\infty} |x_n|^2$ converges. Vector addition in ℓ_2 is the usual vector addition of two sequences, and multiplication by a scalar is also the standard one. We will later prove that ℓ_2 is a linear space over $\mathbb{C}$ (the fact that ℓ_2 is closed under vector addition is not obvious). For each $\mathbf{x}, \mathbf{y} \in \ell_2$ we define

$$\langle \mathbf{x}, \mathbf{y} \rangle = \mathbf{x} \cdot \mathbf{y} = \sum_{n=1}^{\infty} x_n \overline{y_n}.$$

It is easily shown that $\langle \cdot, \cdot \rangle$ satisfies all four conditions in the definition of the inner product. It is more difficult to prove that the series $\sum_{n=1}^{\infty} x_n \overline{y_n}$ converges for every $\mathbf{x}, \mathbf{y} \in \ell_2$. This fact is a consequence of the Cauchy-Schwarz inequality which we prove in the next section.

Exercises

1. Let $V_1, V_2, \ldots, V_n$ be linear subspaces of a linear space U over $\mathbb{C}$. Prove that $V = \bigcap_{k=1}^{n} V_k$ is a linear space. What can you say about $W = \bigcup_{k=1}^{n} V_k$?

2. Prove that the set

   $$V = \{ f \mid f : \mathbb{R} \to \mathbb{R},\ f \text{ is absolutely integrable over } \mathbb{R} \}$$

   is a linear space over $\mathbb{R}$.

3. Let $C[-1, 2]$ denote the space of continuous complex-valued functions $f : [-1, 2] \to \mathbb{C}$. Which of the following define an inner product on $C[-1, 2]$, and which do not? Explain.
   (a) $\langle f, g \rangle = \int_{-1}^{2} |f(t) + g(t)|\, dt$
   (b) $\langle f, g \rangle = \int_{-1}^{2} f(t)\overline{g(t)}\, dt + f(-\tfrac{1}{2})\overline{g(-\tfrac{1}{2})}$
   (c) $\langle f, g \rangle = 3 \int_{-1}^{2} f(t)\overline{g(t)}\, dt$
   (d) $\langle f, g \rangle = f(0)\overline{g(0)} + f(1)\overline{g(1)}$

4. Let $V = C^2[-\pi, \pi]$ be the space of real-valued twice continuously differentiable functions defined on the interval $[-\pi, \pi]$. Set

   $$\langle f, g \rangle = f(-\pi)g(-\pi) + \int_{-\pi}^{\pi} f''(x)g''(x)\, dx.$$

   Is this an inner product on V?

5. Let $C^1[0,1]$ denote the space of continuous functions $f : [0,1] \to \mathbb{C}$ with a continuous first derivative on $[0,1]$. Set

$$\langle f, g \rangle = f(0)\overline{g(0)} + f'(0)\overline{g'(0)} + f(1)\overline{g(1)}.$$

(a) Is this an inner product on the subspace $\mathcal{P}_2 = \text{span}\{1, x, x^2\}$?

(b) Is it an inner product on $C^1[0,1]$?

6. Let $C^1[0,1]$ be as in Exercise 5. Which of the following define an inner product on $C^1[0,1]$?

(a) $\langle f, g \rangle = f(0)\overline{g(0)} + \int_0^1 f'(t)\overline{g'(t)}\, dt$

(b) $\langle f, g \rangle = f(0)\overline{g(0)} + f'(1)\overline{g'(1)}$

(c) $\langle f, g \rangle = 2\int_0^{\frac{1}{2}} f(t)\overline{g(t)}\, dt - \int_{\frac{1}{2}}^1 f(t)\overline{g(t)}\, dt$

2. The Norm

There is a connection between the concept of a norm and that of an inner product. The definition of a norm is in no way dependent upon that of an inner product. However, in every inner product space one can always define a norm in a very natural way.

Definition 1.5: *Let V be a linear space. A norm on V is a function from V to $\mathbb{R}_+$ which we denote by $\|\cdot\|$, and which satisfies the following properties:*

(1) *For each $v \in V$, $\|v\| \geq 0$.*

(2) *$\|v\| = 0$ if and only if $v = \vec{0}$.*

(3) *For each $v \in V$ and $a \in \mathbb{C}$, $\|av\| = |a| \cdot \|v\|$.*

(4) *For every $u, v \in V$, $\|u + v\| \leq \|u\| + \|v\|$ (the triangle inequality).*

The concept of a norm is a generalization of the concept of *size* or *distance* (from the zero vector). For every $u, v \in V$, we may consider the number $\|u - v\|$ as the distance between u and v. Hence $\|u\|$ is the distance of u from $\vec{0}$, or the size of u.

The simplest examples of norms are the absolute value on $\mathbb{R}$ and $\mathbb{C}$. Here are some more:

Example 1.5: If $V = \mathbb{R}^n$ or $V = \mathbb{C}^n$, then for each $\mathbf{x} = (x_1, x_2, \ldots, x_n) \in V$ we define

$$\|\mathbf{x}\| = \sqrt{\sum_{k=1}^n |x_k|^2}.$$

We will soon see that this formula defines a norm on V. It is called the *Euclidean norm*.

Example 1.6: Let V be as in the previous example. We define thereon a different norm. Let
$$\|\mathbf{x}\|_\infty = \max\left\{|x_k| \,\Big|\, k = 1, 2, \ldots, n\right\}.$$
This is called the *uniform norm*.

Example 1.7: On the linear spaces $V = \mathbb{R}^n$ and $V = \mathbb{C}^n$ we can define many different norms. One common norm thereon, other than the previous two, is
$$\|\mathbf{x}\|_1 = \sum_{k=1}^{n} |x_k|.$$
It is easy to check that this is in fact a norm.

Example 1.8: If $V = C[a, b]$ then for each $f \in V$,
$$\|f\|_\infty = \max\left\{|f(x)| \,\Big|\, a \leq x \leq b\right\}$$
is a norm. This norm is also called the *uniform norm*.

If V is an inner product space, then there is a natural norm defined thereon. It is given by
$$\|v\| = \sqrt{\langle v, v \rangle}.$$
To prove that this is indeed a norm on V, we first prove the following important inequality.

Theorem 1.6: **(Cauchy-Schwarz Inequality)** *Let V be an inner product space. For each $u, v \in V$ we have*
$$|\langle u, v \rangle| \leq \|u\| \cdot \|v\|.$$

Proof: If $\langle u, v \rangle = 0$, then there is nothing to prove. So let us assume that $\langle u, v \rangle \neq 0$ (and thus $u, v \neq \vec{0}$). For convenience set $a = \langle u, v \rangle$ (a may be a complex number). Thus for every real number λ we have

$$\begin{aligned}
0 \leq \|u - \lambda a v\|^2 &= \langle u - \lambda a v, u - \lambda a v \rangle \\
&= \langle u, u \rangle - \lambda \langle u, av \rangle - \lambda \langle av, u \rangle + \lambda^2 \langle av, av \rangle \\
&= \|u\|^2 - \lambda \overline{a} \langle u, v \rangle - \lambda a \langle v, u \rangle + \lambda^2 a \overline{a} \langle v, v \rangle \\
&= \|u\|^2 - \lambda \overline{a} a - \lambda a \overline{a} + \lambda^2 a \overline{a} \|v\|^2 \\
&= \|u\|^2 - 2\lambda |a|^2 + \lambda^2 |a|^2 \|v\|^2.
\end{aligned}$$

We consider the last expression as a quadratic polynomial in λ which is non-negative for every real λ. For this to happen the discriminant must be non-positive, from which the result follows. Equivalently, setting $\lambda = \frac{1}{\|v\|^2}$ (the minimal value of the polynomial) in the above expression leads to

$$0 \leq \|u\|^2 \|v\|^2 - |a|^2.$$

Thus $|a| \leq \|u\| \|v\|$. Since $a = \langle u, v \rangle$ this proves the desired inequality. ∎

Theorem 1.7: *Let V be an inner product space. For each $v \in V$ the equation $\|v\| = \sqrt{\langle v, v \rangle}$ defines a norm on V.*

Proof: By the definition of an inner product, the value $\langle v, v \rangle$ is non-negative for every $v \in V$. This being so, the expression $\sqrt{\langle v, v \rangle}$ is well defined (we take the non-negative square root). Now $\|v\| = 0$ if and only if $\langle v, v \rangle = 0$. By the definition of the inner product this condition is equivalent to $v = \vec{0}$. Thus Conditions 1 and 2 are satisfied. To prove Condition 3, let $a \in \mathbb{C}$. Then

$$\|av\|^2 = \langle av, av \rangle = |a|^2 \langle v, v \rangle = |a|^2 \cdot \|v\|^2$$

and thus $\|av\| = |a| \cdot \|v\|$.

It remains to prove the triangle inequality. To this end, let $u, v \in V$. Then

$$\|u + v\|^2 = \langle u + v, u + v \rangle$$
$$= \langle u, u \rangle + \langle u, v \rangle + \langle v, u \rangle + \langle v, v \rangle$$
$$= \|u\|^2 + \langle u, v \rangle + \overline{\langle u, v \rangle} + \|v\|^2.$$

The expression $\langle u, v \rangle + \overline{\langle u, v \rangle}$ is a real number. From the Cauchy-Schwarz inequality it follows that

$$\left| \langle u, v \rangle + \overline{\langle u, v \rangle} \right| \leq 2 |\langle u, v \rangle| \leq 2 \|u\| \cdot \|v\|.$$

Thus

$$\|u + v\|^2 \leq \|u\|^2 + 2\|u\| \cdot \|v\| + \|v\|^2 = (\|u\| + \|v\|)^2.$$

This proves the triangle inequality and the theorem. ∎

Among the various examples we gave of norms, only the Euclidean norm is a norm defined via an inner product. The inner product on which it is based is the standard inner product of Example 1.2. From Theorem 1.7 it therefore follows that the Euclidean norm is really and truly a norm.

From the Cauchy-Schwarz and the triangle inequalities we can derive other more specific inequalities.

Example 1.9: If we consider $\mathbb{R}^n$ with the standard inner product, then for each $\mathbf{x} \in \mathbb{R}^n$ the norm of $\mathbf{x}$ is given by

$$\|\mathbf{x}\| = \sqrt{\mathbf{x} \cdot \mathbf{x}} = \sqrt{\sum_{k=1}^{n} x_k^2}.$$

This is also called the *distance* from $\mathbf{x}$ to $\mathbf{0}$ (recall the Pythagorean Theorem). From the Cauchy-Schwarz inequality we obtain, for every $\mathbf{x}, \mathbf{y} \in \mathbb{R}^n$,

$$\left| \sum_{k=1}^{n} x_k y_k \right| \leq \sqrt{\sum_{k=1}^{n} x_k^2} \sqrt{\sum_{k=1}^{n} y_k^2}.$$

Example 1.10: If we consider $\mathbb{C}^n$ with the standard inner product then as above, the norm of $\mathbf{x} \in \mathbb{C}^n$ is given by

$$\|\mathbf{x}\| = \sqrt{\mathbf{x} \cdot \mathbf{x}} = \sqrt{\sum_{k=1}^{n} |x_k|^2}.$$

From the Cauchy-Schwarz inequality we obtain, for every $\mathbf{x}, \mathbf{y} \in \mathbb{C}^n$,

$$(1.1) \qquad \left| \sum_{k=1}^{n} x_k \overline{y_k} \right| \leq \sqrt{\sum_{k=1}^{n} |x_k|^2} \sqrt{\sum_{k=1}^{n} |y_k|^2}.$$

As a result of the triangle inequality we have, for each $\mathbf{x}, \mathbf{y} \in \mathbb{C}^n$,

$$(1.2) \qquad \sqrt{\sum_{k=1}^{n} |x_k + y_k|^2} \leq \sqrt{\sum_{k=1}^{n} |x_k|^2} + \sqrt{\sum_{k=1}^{n} |y_k|^2}.$$

Example 1.11: We now use the previous results to prove that ℓ_2 is in fact a linear space, and that the inner product defined in Example 1.4 is indeed an inner product. Let $\mathbf{x}, \mathbf{y} \in \ell_2$. We first prove that the inner product $\mathbf{x} \cdot \mathbf{y}$ is well defined. We recall that

$$\mathbf{x} \cdot \mathbf{y} = \sum_{n=1}^{\infty} x_n \overline{y_n} = \lim_{m \to \infty} \sum_{n=1}^{m} x_n \overline{y_n}.$$

We prove that the above series converges absolutely, i.e., the series $\sum_{n=1}^{\infty} |x_n \overline{y_n}|$ converges. For each natural integer m we have from (1.1)

$$\sum_{n=1}^{m} |x_n \overline{y_n}| \leq \sqrt{\sum_{n=1}^{m} |x_n|^2} \sqrt{\sum_{n=1}^{m} |y_n|^2} \leq \sqrt{\sum_{n=1}^{\infty} |x_n|^2} \sqrt{\sum_{n=1}^{\infty} |y_n|^2} = \|\mathbf{x}\| \cdot \|\mathbf{y}\|.$$

The partial sums $S_m = \sum_{n=1}^{m} |x_n \overline{y_n}|$ are bounded above by $\|\mathbf{x}\| \cdot \|\mathbf{y}\|$. Since these partial sums $\{S_m\}_{m=1}^{\infty}$ are monotonically increasing, they must converge. Thus the series $\sum_{n=1}^{\infty} x_n \overline{y_n}$ converges, and the inner product on ℓ_2 is well defined. It remains to prove that ℓ_2 is closed under addition. That is, we show that if $\mathbf{x}, \mathbf{y} \in \ell_2$ then $\mathbf{x} + \mathbf{y} \in \ell_2$. We must verify that for $\mathbf{x}, \mathbf{y} \in \ell_2$ the series $\sum_{n=1}^{\infty} |x_n + y_n|^2$ converges. From the inequality (1.2) we have that for each m

$$\sum_{n=1}^{m} |x_n + y_n|^2 \le \left[\left(\sum_{n=1}^{m} |x_n|^2\right)^{\frac{1}{2}} + \left(\sum_{n=1}^{m} |y_n|^2\right)^{\frac{1}{2}}\right]^2$$
$$\le (\|\mathbf{x}\| + \|\mathbf{y}\|)^2.$$

The partial sums $\sum_{n=1}^{m} |x_n + y_n|^2$ are monotonically increasing and bounded above by the finite value $(\|\mathbf{x}\| + \|\mathbf{y}\|)^2$. Thus the series converges and $\mathbf{x} + \mathbf{y} \in \ell_2$.

Example 1.12: With the inner product of Example 1.3, the Cauchy-Schwarz inequality on the space of continuous functions $C[a,b]$ is given by

$$\left|\int_a^b f(x)\overline{g(x)}\,dx\right|^2 \le \left(\int_a^b |f(x)|^2 dx\right)\left(\int_a^b |g(x)|^2 dx\right).$$

Exercises

1. Prove that the "norms" defined in Examples 1.6–1.8 are in fact norms.

2. (a) Prove that for any $f, g \in C[a,b]$, with inner product
$$\langle f, g \rangle = \int_a^b f(x)\overline{g(x)}\,dx,$$
we have
$$\frac{1}{2}\int_a^b \int_a^b |f(x)g(y) - g(x)f(y)|^2\,dx\,dy = \|f\|^2 \cdot \|g\|^2 - |\langle f, g\rangle|^2.$$
   (b) Use the equality in (a) to prove the Cauchy-Schwarz inequality on $C[a,b]$.

3. Let V be an inner product space, $u, v \in V$, $u, v \ne \vec{0}$.
   (a) Show that $\langle u, v \rangle = \|u\| \cdot \|v\|$ if and only if $u = av$ for some $a \in \mathbb{C}$.
   (b) Show that $\|u + v\| = \|u\| + \|v\|$ if and only if $u = av$ for some $a \ge 0$.

4. Let V be an inner product space. Prove that for all $u, v \in V$ the "Parallelogram law"
$$\|u+v\|^2 + \|u-v\|^2 = 2\|u\|^2 + 2\|v\|^2$$
holds.

5. Let V be a real inner product space. Prove that for each $u, v \in V$
$$\langle u, v \rangle = \frac{1}{4}\|u+v\|^2 - \frac{1}{4}\|u-v\|^2.$$

6. Let V be a complex inner product space. Prove that for each $u, v \in V$

$$\langle u, v \rangle = \frac{1}{4}\|u+v\|^2 - \frac{1}{4}\|u-v\|^2 + \frac{i}{4}\|u+iv\|^2 - \frac{i}{4}\|u-iv\|^2.$$

7. On the basis of the previous two exercises prove that in every normed linear space for which the parallelogram law holds we can define an inner product associated with the given norm.

8. Prove that for any natural number n and real numbers $x_1, x_2, \ldots, x_n$ we have the inequality

$$\left|\frac{1}{n}\sum_{k=1}^{n} x_k\right| \leq \left(\frac{1}{n}\sum_{k=1}^{n} x_k^2\right)^{\frac{1}{2}}.$$

3. Orthogonal and Orthonormal Systems

Definition 1.8: *Let V be an inner product space and $u, v \in V$. We say that u and v are* perpendicular *to each other, or* orthogonal, *if $\langle u, v \rangle = 0$. We denote this fact by $u \perp v$.*

Definition 1.9: *Let V be an inner product space. A finite sequence $\{u_k\}_{k=1}^{n}$ or an infinite sequence $\{u_k\}_{k=1}^{\infty}$ of vectors in V is called an* orthogonal system *if $u_k \neq \vec{0}$ for each k and $u_k \perp u_j$ for all $k \neq j$. If, in addition, $\|u_k\| = 1$ for every k, then we say that we have an* orthonormal system.

Every vector of length 1 (i.e., for which $\|u\| = 1$) is called a *unit vector*. This being so, an orthonormal system is an orthogonal system where each of its vectors is a unit vector. If we are given an orthogonal system $\{u_k\}_{k=1}^{n}$ (where n is finite or infinite) then we can easily obtain an orthonormal system by "normalizing" each vector of the system. For each k, set $e_k = \frac{u_k}{\|u_k\|}$. It follows that $\|e_k\| = 1$ for each k, and $e_k \perp e_j$ for $k \neq j$. Thus $\{e_k\}_{k=1}^{n}$ is an orthonormal system of the same size as the orthogonal system $\{u_k\}_{k=1}^{n}$. More importantly, $\text{span}\{e_k\}_{k=1}^{n} = \text{span}\{u_k\}_{k=1}^{n}$.

Proposition 1.10: *Let $\{u_k\}_{k=1}^{n}$ be a finite orthogonal system in an inner product space V. Then the vectors $\{u_k\}_{k=1}^{n}$ are linearly independent.*

Proof: Assume that

$$a_1 u_1 + a_2 u_2 + \cdots + a_n u_n = \vec{0}, \qquad a_1, a_2, \ldots, a_n \in \mathbb{C}.$$

For each $k \in \{1, 2, \ldots, n\}$

$$\begin{aligned}
0 = \langle \vec{0}, u_k \rangle &= \langle a_1 u_1 + a_2 u_2 + \cdots + a_n u_n, u_k \rangle \\
&= a_1 \langle u_1, u_k \rangle + a_2 \langle u_2, u_k \rangle + \cdots + a_k \langle u_k, u_k \rangle + \cdots + a_n \langle u_n, u_k \rangle \\
&= a_1 \cdot 0 + \cdots + a_{k-1} \cdot 0 + a_k \cdot \langle u_k, u_k \rangle + a_{k+1} \cdot 0 + \cdots + a_n \cdot 0 \\
&= a_k \langle u_k, u_k \rangle.
\end{aligned}$$

Since $\langle u_k, u_k \rangle = \|u_k\|^2 \neq 0$, we must have $a_k = 0$. Thus $a_k = 0$ for each k, and the vectors $\{u_k\}_{k=1}^n$ are linearly independent. ∎

Let $\{v_k\}_{k=1}^n$ be any finite system of linearly independent vectors in an inner product space V. Does there then exist an orthonormal system $\{e_k\}_{k=1}^n$ for which

$$\text{span}\{e_k\}_{k=1}^n = \text{span}\{v_k\}_{k=1}^n \,?$$

And if yes, is there a simple method of constructing such a system $\{e_k\}_{k=1}^n$ from the original system $\{v_k\}_{k=1}^n$? The answer to both questions is yes. One known method of constructing an orthonormal system from the original system is called the *Gram-Schmidt process*. We will quickly sketch this process in the next section.

One of the many advantages of an orthonormal system is the relative ease with which we can determine the coefficients of any vector in its linear span. The formula for the coefficients is to be found in this next proposition.

Proposition 1.11: *Assume V is an inner product space, and $\{e_1, \ldots, e_n\}$ an orthonormal system therein. If $u = \sum_{k=1}^n a_k e_k$ then for each k we have $a_k = \langle u, e_k \rangle$.*

Proof:

$$\begin{aligned}
\langle u, e_k \rangle &= \langle a_1 e_1 + a_2 e_2 + \cdots + a_n e_n, e_k \rangle \\
&= a_1 \langle e_1, e_k \rangle + a_2 \langle e_2, e_k \rangle + \cdots + a_k \langle e_k, e_k \rangle + \cdots + a_n \langle e_n, e_k \rangle \\
&= a_1 \cdot 0 + a_2 \cdot 0 + \cdots + a_k \cdot 1 + \cdots + a_n \cdot 0 \\
&= a_k.
\end{aligned}$$
∎

Thus if $\{e_1, \ldots, e_n\}$ is an orthonormal system, then for each u in its span we have

$$u = \sum_{k=1}^n a_k e_k = \sum_{k=1}^n \langle u, e_k \rangle e_k.$$

The coefficient a_k is uniquely determined by the formula $a_k = \langle u, e_k \rangle$. Note that a_k is only dependent upon u and e_k. It is not dependent upon any of the other basis vectors e_j, $j \neq k$. In general, if $u \in \text{span}\{v_1, \ldots, v_n\}$, where $\{v_1, \ldots, v_n\}$ is not an orthonormal system, then each of the coefficients a_k in the representation of u as a linear combination of $v_1, \cdots, v_n$ will depend upon every one of $v_1, \ldots, v_n$ in some complicated manner. The above coefficients are of such importance that they have a name.

Definition 1.12: *Let V be an inner product space and $\{e_k\}_{k=1}^n$ an orthonormal system therein (n may be finite or infinite). Let $u \in V$. The numbers $\langle u, e_k \rangle$ are called the* **generalized Fourier coefficients** *of the vector u with respect to the given orthonormal system.*

An additional property of an orthonormal system is presented in the next proposition.

Proposition 1.13: *Let V be an inner product space and $\{e_1, \ldots, e_n\}$ an orthonormal system therein. If $\{a_k\}_{k=1}^n$ and $\{b_k\}_{k=1}^n$ are any sequences of scalars, then*

$$\left\langle \sum_{k=1}^n a_k e_k, \sum_{k=1}^n b_k e_k \right\rangle = \sum_{k=1}^n a_k \overline{b_k},$$

i.e., for $u, v \in \text{span}\{e_1, \ldots, e_n\}$,

$$\langle u, v \rangle = \sum_{k=1}^n \langle u, e_k \rangle \overline{\langle v, e_k \rangle}.$$

The proof of this proposition is similar to the proofs of Propositions 1.10 and 1.11. Note that there is also an analog of the last formula for calculating $\langle u, v \rangle$ in the case when our basis is not orthonormal. But it contains n^2 rather than only n terms.

This next theorem may be viewed as a generalization of the Pythagorean Theorem in an inner product space.

Theorem 1.14: **(Generalized Pythagorean Theorem)** *Let V be an inner product space.*

(a) *Let $\{u_1, \ldots, u_n\}$ be an orthogonal system in V, and $a_1, \ldots, a_n$ scalars. Then*

$$\left\| \sum_{k=1}^n a_k u_k \right\|^2 = \sum_{k=1}^n |a_k|^2 \|u_k\|^2.$$

(b) Let $\{e_1, \ldots, e_n\}$ be an orthonormal system in V. Then for every $u \in \mathrm{span}\{e_1, \ldots, e_n\}$
$$\|u\|^2 = \sum_{k=1}^{n} |\langle u, e_k \rangle|^2.$$

Proof: (a) This follows from the definitions.

$$\left\|\sum_{k=1}^{n} a_k u_k\right\|^2 = \left\langle \sum_{k=1}^{n} a_k u_k, \sum_{j=1}^{n} a_j u_j \right\rangle = \sum_{k=1}^{n} \sum_{j=1}^{n} a_k \bar{a}_j \langle u_k, u_j \rangle$$

$$= \sum_{k=1}^{n} |a_k|^2 \|u_k\|^2.$$

(b) This is an immediate consequence of (a) or of Proposition 1.13. ∎

We should consider (b) of the above theorem as a natural generalization of the Euclidean norm which was defined on $\mathbb{R}^n$ and $\mathbb{C}^n$ (see Example 1.5). From the isomorphism which exists between $W = \mathrm{span}\{e_1, \ldots, e_n\}$ and $\mathbb{C}^n$ (here we assume that V is a complex linear space) we can identify each vector $u \in W$ with its sequence of generalized Fourier coefficients $(a_1, a_2, \ldots, a_n) \in \mathbb{C}^n$. Theorem 1.14 effectively says that $\|u\| = \|(a_1, \ldots, a_n)\|$, where the norm on the right-hand side is the Euclidean norm on $\mathbb{C}^n$. Thus there is not, in some sense, a significant difference between W and $\mathbb{C}^n$.

Exercises

1. Let $C[-1, 1]$ be the space of continuous functions $f : [-1, 1] \to \mathbb{C}$ with inner product
$$\langle f, g \rangle = \int_{-1}^{1} f(x) \overline{g(x)} \, dx.$$

   (a) Let $P_0(x) = 1$, $P_1(x) = x$, and $P_2(x) = 1 - 3x^2$. Prove that this set of polynomials is orthogonal in $C[-1, 1]$.

   (b) Find constants a, b, and c such that the polynomial
$$P_3(x) = a + bx + cx^2 + x^3$$
   is orthogonal (perpendicular) to each of the previous polynomials.

2. Let $\mathcal{P}_2$ be the space of real polynomials of degree less than or equal to 2. To each $f, g \in \mathcal{P}_2$, define
$$\langle f, g \rangle = \int_0^\infty f(x) g(x) e^{-x} dx.$$

(a) Prove that this is an inner product on $\mathcal{P}_2$.

(b) Show that the set $\{1,\ 1-x,\ 1-2x+\tfrac{1}{2}x^2\}$ is an orthonormal system with respect to this inner product.

3. Let $C^1[0,1]$ be the space of continuous functions $f : [0,1] \to \mathbb{C}$ with a continuous first derivative on $[0,1]$.

(a) Prove that
$$\langle f,g\rangle = f(0)\cdot\overline{g(0)} + \int_0^1 f'(x)\overline{g'(x)}\,dx$$
is an inner product on $C^1[0,1]$.

(b) Find an orthonormal system $\{h_1, h_2, h_3\}$ in $C^1[0,1]$, with respect to this inner product, for which
$$\text{span}\{h_1,h_2,h_3\} = \text{span}\{1,x,x^2\}.$$

4. Orthogonal Projections and Approximation in the Mean

Let V be an inner product space, and $\{e_1,\ldots,e_n\}$ an orthonormal system therein. Set $W = \text{span}\{e_1,\ldots,e_n\}$. Let u be an arbitrary vector in V. In the previous section we defined the generalized Fourier coefficients of u to be the $\langle u, e_k\rangle$, $k=1,\ldots,n$. If $u \notin W$ then $u \neq \sum_{k=1}^n \langle u,e_k\rangle e_k$, since u is not a linear combination of the $e_1, \ldots, e_n$. Nevertheless there exists an important connection between u and $\sum_{k=1}^n \langle u,e_k\rangle e_k$. In this section we study this link in some detail.

For each $u \in V$, we set $\tilde{u} = \sum_{k=1}^n \langle u,e_k\rangle e_k$. The vector $\tilde{u}$ is said to be the *orthogonal projection of u on W*.

Proposition 1.15: *For each $u \in V$,*
(a) $\langle u - \tilde{u}, w\rangle = 0$ *for all $w \in W$.*
(b) $\|u - w\|^2 = \|u - \tilde{u}\|^2 + \|\tilde{u} - w\|^2$ *for all $w \in W$.*

Remark: Part (a) of this proposition says that the vector $u - \tilde{u}$ is orthogonal to every vector $w \in W$. This being so, we sometimes say that $u - \tilde{u}$ is orthogonal to the subspace W, and write $u - \tilde{u} \perp W$.

Proof: (a) We first prove that $\langle u - \tilde{u}, e_j\rangle = 0$ for every $j = 1, 2, \ldots, n$.

$$\langle u-\tilde{u}, e_j\rangle = \langle u, e_j\rangle - \left\langle \sum_{k=1}^n \langle u,e_k\rangle e_k, e_j\right\rangle = \langle u,e_j\rangle - \sum_{k=1}^n \langle u,e_k\rangle \langle e_k, e_j\rangle$$
$$= \langle u,e_j\rangle - \langle u,e_j\rangle \langle e_j, e_j\rangle = \langle u,e_j\rangle - \langle u,e_j\rangle = 0.$$

We now take an arbitrary $w \in W$. Thus $w = \sum_{j=1}^{n} b_j e_j$ for some scalars $b_1, b_2, \ldots, b_n$, and

$$\langle u - \tilde{u}, w \rangle = \left\langle u - \tilde{u}, \sum_{j=1}^{n} b_j e_j \right\rangle = \sum_{j=1}^{n} \overline{b_j} \langle u - \tilde{u}, e_j \rangle = \sum_{j=1}^{n} \overline{b_j} \cdot 0 = 0.$$

(b)

From part (a) we have $(u - \tilde{u}) \perp w$ for every $w \in W$. Thus $(u - \tilde{u}) \perp (\tilde{u} - w)$ since $\tilde{u} - w \in W$. From part (a) of Theorem 1.14 we obtain

$$\|u - w\|^2 = \|u - \tilde{u} + \tilde{u} - w\|^2 = \|u - \tilde{u}\|^2 + \|\tilde{u} - w\|^2$$

and this proposition is proved. ∎

We present consequences of this last result after the definition of distance (which we have already met in Section 1.2).

Definition 1.16: *Let V be any normed linear space. For each $u, v \in V$, the distance between u and v is the non-negative number $\|u - v\|$.*

To justify the term "distance" we list a number of basic properties which our definition of distance satisfies:

(a) For each $u, v \in V$, we have $\|u - v\| = \|v - u\|$. That is, the distance between u and v equals the distance between v and u.

(b) For each $u \in V$, $\|u - u\| = 0$. That is, u is distant zero from itself.

(c) For every $u, v \in V$ we have $\|u - v\| = 0$ only if $u = v$. That is, if the distance between u and v is zero then u equals v.

(d) For every $u, v, w \in V$, we have $\|u - w\| \le \|u - v\| + \|v - w\|$. That is, the distance between u and w is always less than or equal to the sum of the distances from u and w to any third point v.

Properties (a)–(c) are direct consequences of the definition of a norm. Property (d) follows from the triangle inequality

$$\|u - w\| = \|u - v + v - w\| \le \|u - v\| + \|v - w\|.$$

The main result of this section is the following characterization of the vector closest to u in W.

Theorem 1.17: *Let V be an inner product space and $\{e_1, \ldots, e_n\}$ an orthonormal system therein. Set $W = \mathrm{span}\{e_1, \ldots, e_n\}$, and let $u \in V$. The vector*

$\tilde{u} = \sum_{k=1}^{n} \langle u, e_k \rangle e_k$ *is a closest vector to u in W. In addition, $\tilde{u}$ is the unique vector in W whose distance from u is minimal.*

Proof: We must show that $\|u - \tilde{u}\| \leq \|u - w\|$ for all $w \in W$. This is the meaning of the expression "a closest vector to u from W". This is an immediate consequence of part (b) of Proposition 1.15. For each $w \in W$,

$$\|u - w\|^2 = \|u - \tilde{u}\|^2 + \|\tilde{u} - w\|^2$$

and thus $\|u - \tilde{u}\| \leq \|u - w\|$ for each $w \in W$. The uniqueness of $\tilde{u}$ as the closest vector is a result of this same equality. If $\|u - \tilde{u}\| = \|u - w\|$ for some $w \in W$, then $\|\tilde{u} - w\| = 0$, which implies that $w = \tilde{u}$. ∎

As we see $\tilde{u}$ has a simple affinity to u. It is the unique closest vector to u from W. Naturally if $u \in W$ then $\tilde{u} = u$. We will further consider this relationship when we deal with how $\tilde{u}$ might "represent" u and why.

Example 1.13: For the linear space $C[-1, 1]$, the norm

$$\|f\| = \left(\int_{-1}^{1} |f(x)|^2 \, dx \right)^{\frac{1}{2}}$$

comes from the inner product

$$\langle f, g \rangle = \int_{-1}^{1} f(x) \overline{g(x)} \, dx.$$

In order to determine the closest function to $f \in C[-1, 1]$ in the subspace $W = \text{span}\{1, x\}$ we must find scalars a^* and b^* for which

$$\|f - (a^* + b^* x)\| \leq \|f - (a + bx)\|$$

for all $a, b \in \mathbb{C}$. The functions $e_1(x) = \frac{1}{\sqrt{2}}$ and $e_2(x) = \sqrt{\frac{3}{2}} x$ form an orthonormal basis for the space W. Thus our problem is equivalent to that of finding scalars c^* and d^* for which

$$\|f - (c^* e_1 + d^* e_2)\| \leq \|f - (ce_1 + de_2)\|$$

for all $c, d \in \mathbb{C}$. According to Theorem 1.17 there is a unique solution to this problem and it is given by

$$c^* = \langle f, e_1 \rangle, \qquad d^* = \langle f, e_2 \rangle.$$

If, for example, $f(x) = x^3$ then

$$c^* = \langle f, e_1 \rangle = \int_{-1}^{1} x^3 \frac{1}{\sqrt{2}} \, dx = 0,$$

$$d^* = \langle f, e_2 \rangle = \int_{-1}^{1} x^3 \sqrt{\frac{3}{2}} x \, dx = \frac{\sqrt{6}}{5},$$

and thus
$$c^* e_1(x) + d^* e_2(x) = \frac{\sqrt{6}}{5}\sqrt{\frac{3}{2}}x = \frac{3}{5}x$$
is the closest function to x^3 in W with respect to the given norm.

One additional consequence of Proposition 1.15 is the following.

Proposition 1.18: Let $\{e_1, \ldots, e_n\}$ be an orthonormal system in an inner product space V. Then for each $u \in V$ the following inequality holds:
$$\sum_{k=1}^{n} |\langle u, e_k \rangle|^2 \leq \|u\|^2.$$

Proof: Setting $w = \vec{0}$ in part (b) of Proposition 1.15 we obtain
$$\|u\|^2 = \|u - \tilde{u}\|^2 + \|\tilde{u}\|^2.$$
Thus $\|\tilde{u}\|^2 \leq \|u\|^2$. From part (b) of Theorem 1.14
$$\|\tilde{u}\|^2 = \sum_{k=1}^{n} |\langle u, e_k \rangle|^2$$
and the inequality follows. ∎

It is not difficult to establish that $\sum_{k=1}^{n} |\langle u, e_k \rangle|^2 = \|u\|^2$ if and only if $u \in \text{span}\{e_1, \ldots, e_n\}$.

The Gram-Schmidt Process

Let V be an inner product space and $\{v_1, \ldots, v_n\}$ a system of linearly independent vectors in V. We will describe a method whereby we obtain an orthonormal system $\{e_1, \ldots, e_n\}$ for which
$$\text{span}\{v_1, \ldots, v_n\} = \text{span}\{e_1, \ldots, e_n\}.$$
The process is an n-step method whereby at step k, $1 \leq k \leq n$, we build the vector e_k in such a way that
$$\text{span}\{v_1, \ldots, v_k\} = \text{span}\{e_1, \ldots, e_k\}.$$

Step 1: From the fact that the system $\{v_1, \ldots, v_n\}$ is linearly independent, we have that $v_1 \neq \vec{0}$. We define e_1 by
$$e_1 = \frac{v_1}{\|v_1\|}.$$

Chapter 1: Background

It is clear that $\|e_1\| = 1$ and that $\text{span}\{v_1\} = \text{span}\{e_1\}$.

Step 2: Let $W_1 = \text{span}\{e_1\}$ and let $\tilde{v}_2 = \langle v_2, e_1 \rangle e_1$ be the orthogonal projection of v_2 on W_1. From Proposition 1.15(a), $v_2 - \tilde{v}_2 \perp e_1$. In addition $v_2 - \tilde{v}_2 \neq \vec{0}$ since otherwise we would have $v_2 \in W_1$, contradicting the fact that the system $\{v_1, v_2\}$ is linearly independent. Thus we may define

$$e_2 = \frac{v_2 - \tilde{v}_2}{\|v_2 - \tilde{v}_2\|}.$$

It follows that $e_2 \perp e_1$, $\|e_2\| = 1$, and $\text{span}\{e_1, e_2\} = \text{span}\{v_1, v_2\}$. In other words, the system $\{e_1, e_2\}$ is an orthonormal system and it spans the same subspace as that spanned by $\{v_1, v_2\}$.

Step k: Let $W_{k-1} = \text{span}\{e_1, \ldots, e_{k-1}\}$ and let $\tilde{v}_k = \sum_{j=1}^{k-1} \langle v_k, e_j \rangle e_j$ be the orthogonal projection of v_k on W_{k-1}. From Proposition 1.15(a), $v_k - \tilde{v}_k \perp W_{k-1}$. In addition, $v_k - \tilde{v}_k \neq \vec{0}$. Thus we may define

$$e_k = \frac{v_k - \tilde{v}_k}{\|v_k - \tilde{v}_k\|}.$$

As a consequence $\text{span}\{e_1, \ldots, e_k\} = \text{span}\{v_1, \ldots, v_k\}$, and $\{e_1, \ldots, e_k\}$ is an orthonormal system.

We continue this process until the nth step where we obtain the desired orthonormal system $\{e_1, \ldots, e_n\}$.

Exercises

1. Let $f \in C[-\pi, \pi]$. For each $\alpha, \beta, \gamma \in \mathbb{C}$ define

$$F(\alpha, \beta, \gamma) = \frac{1}{\pi} \int_{-\pi}^{\pi} |f(x) - \alpha - \beta \cos x - \gamma \cos 10x|^2 dx.$$

Prove that F attains its minimum at a unique point $(\alpha_0, \beta_0, \gamma_0)$ and find this point when

(a) $f(x) = \cos^2 x$ (b) $f(x) = x^3$ (c) $f(x) = \sin x$

(d) $f(x) = 1 - 2\cos x$ (e) $f(x) = |x|$ (f) $f(x) = |\sin x|$

2. On the linear space $C[0, 2\pi]$ we define the inner product

$$\langle f, g \rangle = \int_0^{2\pi} f(x) \overline{g(x)} \, dx.$$

(a) Prove that the set $S = \left\{ \frac{1}{\sqrt{2\pi}}, \frac{\cos 2x}{\sqrt{\pi}}, \frac{\sin x}{\sqrt{\pi}} \right\}$ is an orthonormal system.

(b) Let W be the subspace spanned by S, and let $f(x) = x$ on the interval $[0, 2\pi]$. Find the function g in S which is closest to f (that is to say, for which $\|f - g\|$ is minimal).

3. Consider the space $C[-\pi, \pi]$ with the inner product

$$\langle f, g \rangle = \frac{1}{\pi} \int_{-\pi}^{\pi} f(x)\overline{g(x)}\, dx.$$

Let $W = \text{span}\{1, \sin x, \cos x, x\}$, and let $f(x) = |x|$. Find the function $g \in W$ for which $\|f - g\|$ is as small as possible.

4. For each $\alpha, \beta, \gamma \in \mathbb{C}$ define

$$F(\alpha, \beta, \gamma) = \frac{1}{\pi} \int_{-\pi}^{\pi} |x - \alpha - \beta \cos x - \gamma \sin 2x|^2 dx.$$

Show that F attains its minimum at exactly one point $(\alpha_0, \beta_0, \gamma_0)$, and find this point.

5. In the space $C[-1, 1]$ we are given the functions

$$f_0(x) = 1, \ f_1(x) = x + a, \ f_2(x) = x^2 + bx + c, \ f_3(x) = x^3 + Ax^2 + Bx + C$$

and it is known that $\{f_0, f_1, f_2, f_3\}$ is an orthonormal system in $C[-1, 1]$ with respect to the inner product

$$\langle f, g \rangle = \int_{-1}^{1} f(x)\overline{g(x)}\, dx.$$

(a) Calculate a, b, c.

(b) For each $\alpha, \beta, \gamma, \delta \in \mathbb{C}$ define

$$F(\alpha, \beta, \gamma, \delta) = \int_{-1}^{1} |x^4 - \alpha f_0(x) - \beta f_1(x) - \gamma f_2(x) - \delta f_3(x)|^2 dx.$$

Prove that F attains its minimum at exactly one point $(\alpha_0, \beta_0, \gamma_0, \delta_0)$, and find this point.

5. Infinite Orthonormal Systems

Let V be an inner product space. In this section we assume that $\dim(V) = \infty$. Let $\{e_1, e_2, \ldots\}$ be an orthonormal system with an infinite number of vectors. We remark that the concept of a basis in an infinite dimensional linear space is problematic, to say the least. This being so, we should not assume a priori that the given orthonormal system is a basis or even a spanning set for V. We will later return to this troublesome problem.

Theorem 1.19: **(Bessel's Inequality)** *For each $u \in V$, the series $\sum_{n=1}^{\infty} |\langle u, e_n \rangle|^2$ converges. In addition, the inequality*

$$\sum_{n=1}^{\infty} |\langle u, e_n \rangle|^2 \leq \|u\|^2$$

holds.

Proof: This result is an immediate consequence of Proposition 1.18. For each m,

$$S_m = \sum_{n=1}^{m} |\langle u, e_n \rangle|^2 \leq \|u\|^2.$$

That is, the sequence of partial sums $\{S_m\}_{m=1}^{\infty}$ is bounded above by $\|u\|^2$. Since $\{S_m\}_{m=1}^{\infty}$ is a monotonically increasing sequence, it converges to a finite sum. Thus

$$\lim_{m \to \infty} \sum_{n=1}^{m} |\langle u, e_n \rangle|^2 \leq \|u\|^2. \qquad\blacksquare$$

If we have the equality $\sum_{n=1}^{\infty} |\langle u, e_n \rangle|^2 = \|u\|^2$ then we say that *Parseval's identity* (sometimes called *Parseval's equality*) holds for u. An immediate consequence of Bessel's inequality is

Theorem 1.20: **(Riemann-Lebesgue Lemma)** *Let $\{e_1, e_2, \ldots\}$ be an infinite orthonormal system in an inner product space V. Let $u \in V$. Then*

$$\lim_{n \to \infty} \langle u, e_n \rangle = 0.$$

Proof: From Bessel's inequality, the series $\sum_{n=1}^{\infty} |\langle u, e_n \rangle|^2$ converges. If a series converges then the nth coefficient tends to zero as n tends to infinity. $\blacksquare$

As stated at the beginning of this section, the concept of a basis for an infinite dimensional linear space is problematic. This whole question must be treated carefully. One of the first problems which arises has to do with the proper definition of an "infinite linear combination". Let us be more specific. We are given an inner product space V, an infinite sequence of vectors $\{u_1, u_2, \ldots\}$ therein, and a sequence of scalars $\{a_n\}_{n=1}^{\infty}$. Can we give any meaning to the expression "$\sum_{n=1}^{\infty} a_n u_n$"? We are talking about the infinite sum of vectors

$$a_1 u_1 + a_2 u_2 + \cdots + a_n u_n + \cdots$$

and not an infinite sum of numbers! Even if we succeed in giving some meaning to the infinite sum of vectors, it will also be necessary to check that with this meaning a number of properties of a basis are preserved. To this end, we make use of the concept of convergence in norm.

Definition 1.21: *Let $\{w_m\}_{m=1}^{\infty}$ be an infinite sequence of vectors in a normed linear space V. We say that the sequence converges in norm to the vector $w \in V$ if*
$$\lim_{m \to \infty} \|w - w_m\| = 0.$$
That is to say, for each $\epsilon > 0$ there exists an $m(\epsilon)$ such that for all $m \geq m(\epsilon)$ we have $\|w - w_m\| < \epsilon$.

We can now give some meaning to what we mean by an "infinite linear combination" of vectors.

Definition 1.22: *Let $\{u_1, u_2, \ldots\}$ be an infinite sequence of vectors in the normed linear space V. Let $\{a_n\}_{n=1}^{\infty}$ be a sequence of scalars. We say that the series $\sum_{n=1}^{\infty} a_n u_n$ converges in norm to the vector $w \in V$, and write $w = \sum_{n=1}^{\infty} a_n u_n$, if the partial sums $w_m = \sum_{n=1}^{m} a_n u_n$ converge in norm to w. In other words, the series $\sum_{n=1}^{\infty} a_n u_n$ converges in norm to w if*
$$\lim_{m \to \infty} \left\| w - \sum_{n=1}^{m} a_n u_n \right\| = 0.$$

The interpretation of the expression "the vector w is contained in the span of the infinite sequence $\{u_1, u_2, \ldots\}$" is that there exists a sequence of scalars $\{a_n\}_{n=1}^{\infty}$ such that the linear combination $a_1 u_1 + \cdots + a_m u_m$ approaches w, as m grows. The "nearness" of vectors, in a linear space with a norm, is measured by the distance between them.

We can now formulate further desired properties of infinite orthonormal systems.

Definition 1.23: *Let $\{e_1, e_2, \ldots\}$ be an infinite orthonormal system in an inner product space V. We will say that the system is closed in V if for every $u \in V$ we have*
$$\lim_{m \to \infty} \left\| u - \sum_{n=1}^{m} \langle u, e_n \rangle e_n \right\| = 0.$$

Recall that $\sum_{n=1}^{m} \langle u, e_n \rangle e_n$ is the vector closest to u in span$\{e_1, \ldots, e_m\}$. Thus a system is closed in V if for each element of V there is some infinite linear combination from the system which converges in norm to the element.

Defining a property does not by itself imply that the property holds. We will have to work hard in the next chapter to prove this property in the important example considered therein. Meanwhile, we formulate an additional property, equivalent to the property of closedness.

Proposition 1.24: *The infinite orthonormal system $\{e_1, e_2, \ldots\}$ is closed in the inner product space V if and only if for every $u \in V$ we have the equality*

$$\sum_{n=1}^{\infty} |\langle u, e_n \rangle|^2 = \|u\|^2.$$

That is, closedness is equivalent to Parseval's identity holding for all $u \in V$.

Proof: For each natural number m, set $V_m = \text{span}\{e_1, \ldots, e_m\}$, and let $u_m = \sum_{n=1}^{m} \langle u, e_n \rangle e_n$ be the orthogonal projection of u on V_m. From Proposition 1.15(b) (by setting $w = \vec{0}$), we obtain

$$\|u\|^2 = \|u - u_m\|^2 + \|u_m\|^2.$$

Thus

$$\left\| u - \sum_{n=1}^{m} \langle u, e_n \rangle e_n \right\|^2 = \|u\|^2 - \sum_{n=1}^{m} |\langle u, e_n \rangle|^2.$$

This being so, the claim

$$\lim_{m \to \infty} \left\| u - \sum_{n=1}^{m} \langle u, e_n \rangle e_n \right\|^2 = 0$$

is equivalent to

$$\lim_{m \to \infty} \left(\|u\|^2 - \sum_{n=1}^{m} |\langle u, e_n \rangle|^2 \right) = 0.$$

Clearly this equality is equivalent to Parseval's identity $\|u\|^2 = \sum_{n=1}^{\infty} |\langle u, e_n \rangle|^2$. ∎

Another important concept related to orthonormal systems is that of *completeness*.

Definition 1.25: *Let $\{e_1, e_2, \ldots\}$ be an infinite orthonormal system in an inner product space V. We say that the system is* complete *in V if only the zero vector ($u = \vec{0}$) satisfies the equation*

(1.3) $\qquad \langle u, e_n \rangle = 0, \quad n \in \mathbb{N}.$

The property of completeness is certainly one which we would want for any "basis". Note that it is easy to construct an infinite orthonormal system which is neither closed nor complete. We simply take any infinite orthonormal system and remove one of its terms. As we now demonstrate, completeness is weaker than closedness (and it is not equivalent to it).

Proposition 1.26: *If $\{e_1, e_2, \ldots\}$ is a closed infinite orthonormal system in V, then it is also complete in V.*

Proof: Assume that our system is closed in V. Let $u \in V$ be a vector satisfying (1.3). Then by Proposition 1.24
$$\|u\|^2 = \sum_{n=1}^{\infty} |\langle u, e_n \rangle|^2 = \sum_{n=1}^{\infty} |0|^2 = 0$$
and thus we necessarily have $u = \vec{0}$. ∎

If the orthonormal system $\{e_1, e_2, \ldots\}$ is not closed, then for some $u \in V$, the sum $\sum_{n=1}^{m} \langle u, e_n \rangle e_n$ cannot be arbitrarily close to u, for any m. Even if the orthonormal system is closed, the rate at which the above sum approaches u might be insufficient for specific purposes. It might then be necessary to try to find a "better" orthonormal system.

We end this chapter with a generalization of Parseval's identity.

Proposition 1.27: **(Generalized Parseval Identity)** *Let V be an inner product space and $\{e_1, e_2, \ldots\}$ a closed infinite orthonormal system on V. Let $u, v \in V$ and $a_n = \langle u, e_n \rangle$, $b_n = \langle v, e_n \rangle$. Then*
$$\langle u, v \rangle = \sum_{n=1}^{\infty} a_n \overline{b_n}.$$

Proof: We use the equality
$$(1.4) \quad \langle u, v \rangle = \frac{1}{4}\|u + v\|^2 - \frac{1}{4}\|u - v\|^2 + \frac{i}{4}\|u + iv\|^2 - \frac{i}{4}\|u - iv\|^2$$
(see Exercise 6 of Section 1.2). As the orthonormal system $\{e_1, e_2, \ldots\}$ is closed in V, the following four identities follow from Proposition 1.24:

$$\|u + v\|^2 = \sum_{n=1}^{\infty} |a_n + b_n|^2,$$
$$\|u - v\|^2 = \sum_{n=1}^{\infty} |a_n - b_n|^2,$$
$$\|u + iv\|^2 = \sum_{n=1}^{\infty} |a_n + ib_n|^2,$$
$$\|u - iv\|^2 = \sum_{n=1}^{\infty} |a_n - ib_n|^2.$$

By using (1.4) and the identity
$$a_n \overline{b_n} = \frac{1}{4}|a_n + b_n|^2 - \frac{1}{4}|a_n - b_n|^2 + \frac{i}{4}|a_n + ib_n|^2 - \frac{i}{4}|a_n - ib_n|^2$$
we obtain the desired result. ∎

Exercises

1. Let $\{u_n\}_{n=1}^{\infty}$ be an infinite orthogonal system in the inner product space V. Let $v \in V$ satisfy
$$v = \sum_{n=1}^{\infty} a_n u_n.$$
Prove that for each n,
$$a_n = \frac{1}{\|u_n\|^2} \langle v, u_n \rangle.$$

2. For each $n \in \mathbb{N}$, we define the function $f_n : [0,1] \to \mathbb{R}$ by
$$f_n(x) = \begin{cases} 0, & 0 \leq x \leq \frac{1}{n}, \\ \sqrt{n}, & \frac{1}{n} < x < \frac{2}{n}, \\ 0, & \frac{2}{n} \leq x \leq 1 \end{cases}.$$
Prove that the sequence of functions $\{f_n\}_{n=1}^{\infty}$ converges pointwise to the zero function on $[0,1]$, but does not converge to the zero function in the norm
$$\|f\| = \left(\int_0^1 |f(x)|^2 dx \right)^{\frac{1}{2}}.$$

3. Find a sequence of functions in the space $C[0, \infty)$ which converges uniformly to the zero function on the interval $[0, \infty)$, but does not converge to the zero function in the norm
$$\|f\| = \left(\int_0^{\infty} |f(x)|^2 dx \right)^{\frac{1}{2}}.$$

4. Let V be an inner product space and $\{e_n\}_{n=1}^{\infty}$ a closed infinite orthonormal system in V. Let
$$f_{2n-1} = \frac{e_{2n} - e_{2n-1}}{\sqrt{2}}, \quad f_{2n} = \frac{e_{2n} + e_{2n-1}}{\sqrt{2}}, \quad n = 1, 2, 3, \ldots.$$
Prove that $\{f_n\}_{n=1}^{\infty}$ is also a closed infinite orthonormal system in V.

5. Let V be an inner product space.
   (a) Prove that for all $u, v \in V$, $\Big| \|u\| - \|v\| \Big| \leq \|u - v\|$.
   (b) Let $\{u_n\}_{n=1}^{\infty}$ be a sequence of vectors in V which converges in norm to the vector $u \in V$ (i.e., $\lim_{n \to \infty} \|u - u_n\| = 0$). Prove that $\lim_{n \to \infty} \|u_n\| = \|u\|$.

Review Exercises

1. Let $\mathcal{D} = \left\{(x,y) \in \mathbb{R}^2 \mid x^2 + y^2 < 1\right\}$ denote the open unit disk in $\mathbb{R}^2$. Let $C(\mathcal{D})$ be the collection of all complex-valued continuous functions $f : \mathcal{D} \to \mathbb{C}$ defined on $\mathcal{D}$. For each $f, g \in C(\mathcal{D})$ set
$$\langle f, g \rangle = \iint_{\mathcal{D}} f(x,y)\overline{g(x,y)}\, dx\, dy.$$
For each $n \in \mathbb{N}$ we define
$$f_n(x,y) = (x + iy)^n.$$

   (a) Prove that $\langle f, g \rangle$ is an inner product on $C(\mathcal{D})$.

   (b) Prove that the sequence of functions $\{f_n\}_{n=0}^{\infty}$ is an infinite orthogonal system on $C(\mathcal{D})$.

2. For each $f, g \in C[-1, 1]$ we define
$$\langle f, g \rangle = \int_{-1}^{1} \frac{f(x)\overline{g(x)}}{\sqrt{1-x^2}}\, dx.$$

   (a) Prove that this is an inner product on $C[-1, 1]$.

   (b) For each non-negative integer n, define the function T_n in $C[-1, 1]$ by
$$T_n(x) = \cos(n \arccos x).$$
The function T_n is called the *Chebyshev polynomial* of degree n. Prove that the sequence of functions $\{T_n\}_{n=1}^{\infty}$ is an infinite orthonormal system on $C[-1, 1]$.

   (c) By substituting $\theta = \arccos x$, prove that for each n, $T_n(x)$ is in fact a polynomial of degree n in x.

3. The *Legendre polynomials* are defined by
$$P_n(x) = \frac{1}{2^n n!} \frac{d^n}{dx^n}(x^2 - 1)^n.$$

   (a) Prove that
$$\int_{-1}^{1} P_m(x) P_n(x)\, dx = \begin{cases} 0, & m \neq n, \\ \frac{2}{2n+1}, & m = n, \end{cases}$$

and therefore $\{P_n\}_{n=1}^{\infty}$ is an infinite orthogonal system on the space $C[-1, 1]$ with inner product

$$\langle f, g \rangle = \int_{-1}^{1} f(x)\overline{g(x)}\, dx.$$

(b) Let

$$f(x) = \begin{cases} 0, & -1 \leq x \leq 0, \\ 1, & 0 < x \leq 1. \end{cases}$$

Express f as a linear combination of Legendre polynomials.

4. Let N be a natural number. For each $m = 0, 1, \ldots, N-1$, we define a vector in $\mathbb{C}^N$ by

$$u_m = \left(1, e^{\frac{-2\pi i m}{N}}, e^{\frac{-4\pi i m}{N}}, \ldots, e^{\frac{-2(N-1)\pi i m}{N}}\right).$$

Prove that:

(a) The system of vectors $\{u_m\}_{m=0}^{N-1}$ is an orthogonal system in $\mathbb{C}^N$ with the standard inner product.

(b) For each $m = 0, 1, \ldots, N-1$, $\|u_m\|^2 = N$.

(c) For every $u \in \mathbb{C}^N$, we have $u = \frac{1}{N} \sum_{m=0}^{N-1} \langle u, u_m \rangle u_m$.

Remark: The vector of coefficients $\{\langle u, u_m \rangle\}_{m=0}^{N-1}$ is called the *discrete Fourier transform* of u.

Chapter 2
Fourier Series

0. Introduction

In this chapter we study Fourier series. We use Fourier series to represent or approximate functions defined on a finite interval. In this sense Fourier series are similar to polynomials or power series. However, Fourier series are in other ways both better and more general. Fourier series are one example of a closed infinite orthonormal system in an inner product space. They are an application of the general theory presented in the previous chapter. Fourier series also have various specific properties of their own and we shall study some of them. Fourier series were first defined, not too surprisingly, by Jean Baptiste Joseph Fourier (1768–1830) about 200 years ago. That they are an "old" topic does not detract from their importance. Fourier was a mathematician and an engineer who developed these series in order to solve certain problems in partial differential equations. In the last section of this chapter, we present one application of this kind. (Fourier was a participant in the French Revolution. He was with Napoleon in the Egyptian campaign of 1798 and was considered one of the "savants" who accompanied Napoleon in this campaign. He was, for a time, governor of lower Egypt, and later Prefect of Isère (at Grenoble).)

1. Definitions

We will use E to denote the linear space of complex-valued piecewise continuous functions defined on the interval $[-\pi, \pi]$. We recall that a function $f : [-\pi, \pi] \to \mathbb{C}$ is called *piecewise continuous* if it has at most a finite number of points of discontinuity, and in addition, the one-sided limits exist at each

Chapter 2: Fourier Series

point of discontinuity (i.e., the left and right limits exist and are finite). We note that we allow for the possibility that the function may not be defined at a "point of discontinuity". Every point at which the one-sided limits are not equal is called a *jump point* of f. As may be easily seen, sums (and products) of two piecewise continuous functions are themselves piecewise continuous functions. Multiplying a piecewise continuous function by a scalar also gives us a piecewise continuous function. We regard two functions in E as being identical if they are equal at all but a finite number of points. With this convention E is a linear space. Every complex-valued piecewise continuous function defined on $[-\pi, \pi]$ may be written in the form

$$f = u + iv$$

where u and v are real-valued piecewise continuous functions defined on $[-\pi, \pi]$. The function u is called the *real part* of f and we write $u = \mathrm{Re}(f)$. The function v is called the *imaginary part* of f and we write $v = \mathrm{Im}(f)$.

For each $f, g \in E$ we define

(2.1) $$\langle f, g \rangle = \frac{1}{\pi} \int_{-\pi}^{\pi} f(x)\overline{g(x)}\,dx.$$

This integral exists since $f \cdot \bar{g}$ is piecewise continuous. From the basic properties of the Riemann integral it follows that this is an inner product on E. Thus E is an inner product space. Below is a closed infinite orthonormal system on E. In this next theorem we prove the orthonormality.

Theorem 2.1: *The sequence of functions*

$$\left\{ \frac{1}{\sqrt{2}},\ \sin x,\ \cos x,\ \sin 2x,\ \cos 2x,\ \sin 3x,\ \cos 3x,\ \ldots \right\}$$

is an infinite orthonormal system in the space E.

Proof: We must prove that every two distinct functions are orthogonal, and that each function has norm 1. We start with the second of these facts:

$$\left\| \frac{1}{\sqrt{2}} \right\|^2 = \frac{1}{\pi} \int_{-\pi}^{\pi} \left| \frac{1}{\sqrt{2}} \right|^2 dx = \frac{1}{2\pi} \int_{-\pi}^{\pi} dx = 1$$

and for each $n \in \mathbb{N}$,

$$\| \sin nx \|^2 = \frac{1}{\pi} \int_{-\pi}^{\pi} |\sin nx|^2 dx = \frac{1}{\pi} \int_{-\pi}^{\pi} \frac{1 - \cos 2nx}{2}\, dx = 1,$$

$$\| \cos nx \|^2 = \frac{1}{\pi} \int_{-\pi}^{\pi} |\cos nx|^2 dx = \frac{1}{\pi} \int_{-\pi}^{\pi} \frac{1 + \cos 2nx}{2}\, dx = 1.$$

To prove the orthogonality of every two distinct functions in the sequence we must verify the following five claims:

(a) $\left\langle \frac{1}{\sqrt{2}}, \sin nx \right\rangle = 0, \quad n \in \mathbb{N};$

(b) $\left\langle \frac{1}{\sqrt{2}}, \cos nx \right\rangle = 0, \quad n \in \mathbb{N};$

(c) $\langle \sin mx, \cos nx \rangle = 0, \quad m, n \in \mathbb{N};$

(d) $\langle \sin mx, \sin nx \rangle = 0, \quad m \neq n, m, n \in \mathbb{N};$

(e) $\langle \cos mx, \cos nx \rangle = 0, \quad m \neq n, m, n \in \mathbb{N}.$

Each of these claims is a simple exercise. We present the proofs of two of them. We start with the proof of (a).

$$\left\langle \frac{1}{\sqrt{2}}, \sin nx \right\rangle = \frac{1}{\pi} \int_{-\pi}^{\pi} \frac{1}{\sqrt{2}} \sin nx \, dx = \frac{1}{\pi\sqrt{2}} \left[-\frac{\cos nx}{n} \bigg|_{-\pi}^{\pi} \right]$$

$$= \frac{-\cos n\pi + \cos n(-\pi)}{n\pi\sqrt{2}}$$

$$= 0.$$

We now prove (e). For each $m, n \in \mathbb{N}, m \neq n$,

$$\langle \cos mx, \cos nx \rangle = \frac{1}{\pi} \int_{-\pi}^{\pi} \cos mx \cos nx \, dx$$

$$= \frac{1}{\pi} \int_{-\pi}^{\pi} \frac{\cos(m+n)x + \cos(m-n)x}{2} \, dx$$

$$= \frac{1}{2\pi} \left[\frac{\sin(m+n)x}{m+n} \bigg|_{-\pi}^{\pi} + \frac{\sin(m-n)x}{m-n} \bigg|_{-\pi}^{\pi} \right]$$

$$= 0.$$

The other claims are analogously proved. ∎

An important property of the above system is that it is closed in E. This fact would be difficult to prove at this moment. We defer the proof to Section 2.6. If the system were not closed in E, then it would not be so important and useful a system.

In the previous chapter we saw that if $\{e_n\}_{n=1}^{\infty}$ is an orthonormal system, then we can represent each f in the space as an "infinite linear combination" of the form

$$\sum_{n=1}^{\infty} \langle f, e_n \rangle e_n.$$

Let us calculate the terms $\langle f, e_n \rangle e_n$ of this particular series.

Chapter 2: Fourier Series

1. If $e_n(x) = \frac{1}{\sqrt{2}}$, then

$$\langle f, e_n \rangle e_n = \frac{1}{\pi}\left(\int_{-\pi}^{\pi} f(t) \frac{1}{\sqrt{2}}\, dt\right)\frac{1}{\sqrt{2}} = \frac{1}{2\pi}\int_{-\pi}^{\pi} f(t)\, dt.$$

2. If $e_n(x) = \sin nx$, for $n = 1, 2, 3, \ldots$, then

$$\langle f, e_n \rangle e_n = \frac{1}{\pi}\left(\int_{-\pi}^{\pi} f(t) \sin nt\, dt\right) \sin nx.$$

3. If $e_n(x) = \cos nx$, for $n = 1, 2, 3, \ldots$, then

$$\langle f, e_n \rangle e_n = \frac{1}{\pi}\left(\int_{-\pi}^{\pi} f(t) \cos nt\, dt\right) \cos nx.$$

Thus the series $\sum_{n=1}^{\infty} \langle f, e_n \rangle e_n$ for our particular system has the form

(2.2) $$\frac{a_0}{2} + \sum_{n=1}^{\infty} [a_n \cos nx + b_n \sin nx]$$

where

(2.3) $$\begin{cases} a_n = \dfrac{1}{\pi}\int_{-\pi}^{\pi} f(x) \cos nx\, dx, & n = 0, 1, 2, \ldots, \\[1ex] b_n = \dfrac{1}{\pi}\int_{-\pi}^{\pi} f(x) \sin nx\, dx, & n = 1, 2, \ldots. \end{cases}$$

We now formally name and define this series.

Definition 2.2: Let $f \in E$. The series (2.2) associated with f, where a_n and b_n are defined in (2.3), is called the Fourier series of f and we write

$$f(x) \sim \frac{a_0}{2} + \sum_{n=1}^{\infty} [a_n \cos nx + b_n \sin nx].$$

Remarks: (a) Writing the free constant term of the series in the form $\frac{a_0}{2}$ is for convenience and is standard notation. The definition of a_0 is now part of the general definition of all the a_n in (2.3) (since $\cos 0 = 1$). However, this does sometimes confuse the student. Note carefully that $a_0 = \frac{1}{\pi}\int_{-\pi}^{\pi} f(x)\, dx$, but that $\frac{a_0}{2}$ appears in the series.

(b) In the definition of the Fourier series of f we wrote $\sim$ and not equality. There is a reason for this. There is no necessity that the series in question converges for all $x \in [-\pi, \pi]$. And even if the series does converge, it need not converge to the value $f(x)$. This fact is even true if f is continuous on the full interval $[-\pi, \pi]$. We need additional conditions on the function f to ensure that the series converges to the desired values, and in order to obtain the particular type of convergence desired (such as uniform or pointwise convergence).

(c) The Fourier series of f is totally determined by the values of the coefficients a_n and b_n (of which there are a countable number). These coefficients are themselves determined by the specific integrals in (2.3). If we alter the value of the function f at a finite number of points, then the integrals defining a_n and b_n are unchanged. Thus every two functions of E which differ at a finite number of points have exactly the same Fourier series. For this reason, and others, we regard functions which differ at a finite number of points as being identical.

(d) The series (2.2), if it converges for each x, defines a 2π-periodic function on all of $\mathbb{R}$. That is, it satisfies $g(x + 2\pi) = g(x)$ for all $x \in \mathbb{R}$. This is a consequence of this property holding for the functions 1, $\sin nx$, and $\cos nx$, for all $n \in \mathbb{N}$. This being so, it is sometimes convenient to think of each $f \in E$ as a 2π-periodic function defined on all of $\mathbb{R}$, or alternatively as a function defined on the boundary of the unit circle, where x denotes the appropriate angle subtended by the unit circle.

(e) The norm defined by the inner product (2.1) is

$$\|f\| = \left(\frac{1}{\pi} \int_{-\pi}^{\pi} |f(x)|^2 dx \right)^{\frac{1}{2}}.$$

This norm on E has the property that if $\|f\| = 0$ then f is identically zero except for its value on a finite number of points. Once again, we regard as equivalent all functions on E which differ at only a finite number of points (see (c)).

From the definition of the Fourier series we see that the action of taking a function and defining its Fourier series is a linear process. That is to say, if $f, g \in E$, then the Fourier series of $f + g$ is simply the Fourier series of f plus the Fourier series of g (in the sense of summing the coefficients). Similarly, if $a \in \mathbb{C}$, then the Fourier series of af equals a times the Fourier series of f. That is, the Fourier series of af is obtained by simply multiplying each Fourier coefficient of f by a. These two important properties are immediate consequences of basic facts concerning definite integrals.

Before presenting various properties of Fourier series, let us first calculate the Fourier series of a number of elementary functions.

Example 2.1: Let $f(x) = x$ on the interval $[-\pi, \pi]$. Obviously $f \in E$. Let us determine the Fourier series of f. We note that as in many cases, it is worth while (and in some cases necessary) to separately calculate a_0 and a_n ($n \geq 1$).

$$a_0 = \frac{1}{\pi} \int_{-\pi}^{\pi} x \, dx = 0.$$

For $n \geq 1$,

$$a_n = \frac{1}{\pi}\int_{-\pi}^{\pi} x\cos nx\, dx = \frac{1}{\pi}\left[\frac{x\sin nx}{n}\bigg|_{-\pi}^{\pi} - \int_{-\pi}^{\pi}\frac{\sin nx}{n}dx\right]$$

$$= \frac{1}{\pi}\left[\frac{x\sin nx}{n}\bigg|_{-\pi}^{\pi} + \frac{\cos nx}{n^2}\bigg|_{-\pi}^{\pi}\right] = 0.$$

We could have immediately obtained these two results from the oddness of the functions x and $x\cos nx$ on the interval $[-\pi, \pi]$. We will discuss this topic in Section 2.2. We now calculate the b_n, $n \geq 1$.

$$b_n = \frac{1}{\pi}\int_{-\pi}^{\pi} x\sin nx\, dx = \frac{1}{\pi}\left[-\frac{x\cos nx}{n}\bigg|_{-\pi}^{\pi} + \int_{-\pi}^{\pi}\frac{\cos nx}{n}dx\right]$$

$$= \frac{1}{\pi}\left[-\frac{x\cos nx}{n}\bigg|_{-\pi}^{\pi} + \frac{\sin nx}{n^2}\bigg|_{-\pi}^{\pi}\right] = \frac{-2\pi\cos n\pi}{n\pi}$$

$$= \frac{2(-1)^{n+1}}{n}.$$

Thus the Fourier series of $f(x) = x$ is given by

$$x \sim \sum_{n=1}^{\infty} \frac{2(-1)^{n+1}}{n}\sin nx.$$

Already with this simple example we see that the Fourier series cannot agree with the function at every point in $[-\pi, \pi]$. At the endpoints $x = \pm\pi$ the series is equal to zero, and yet the function does not vanish at either endpoint. However, as we will prove, the series does converge to the value x at every interior point of the interval.

Example 2.2: The Fourier series of the function $f(x) = 5 - 3\sin 2x + 8\cos 3x$ is itself (and thus is a finite series in this case). One can obtain this result by calculating the a_n and b_n, but this is totally unnecessary (why??). In general, every function of the form

$$f(x) = \frac{A_0}{2} + \sum_{n=1}^{M} A_n\cos nx + \sum_{n=1}^{M} B_n\sin nx$$

where M is a natural number and $A_M \neq 0$ or $B_M \neq 0$, is called a *trigonometric polynomial of degree M*. For each such function, its Fourier series is itself. That is, the coefficients are

$$a_n = \begin{cases} A_n, & n \leq M, \\ 0, & n > M, \end{cases} \qquad b_n = \begin{cases} B_n, & n \leq M, \\ 0, & n > M. \end{cases}$$

Example 2.3: Let $f(x) = |x|$ on the interval $[-\pi, \pi]$. Obviously $f \in E$. We calculate its Fourier series, and start with a_0.

$$a_0 = \frac{1}{\pi} \int_{-\pi}^{\pi} |x|\, dx = \frac{2}{\pi} \int_0^{\pi} x\, dx = \frac{2}{\pi} \frac{\pi^2}{2} = \pi.$$

For $n \geq 1$

$$a_n = \frac{1}{\pi} \int_{-\pi}^{\pi} |x| \cos nx\, dx = \frac{2}{\pi} \int_0^{\pi} x \cos nx\, dx$$

$$= \frac{2}{\pi} \left[\frac{x \sin nx}{n} \bigg|_0^{\pi} - \int_0^{\pi} \frac{\sin nx}{n} dx \right]$$

$$= \frac{2}{\pi} \left[\frac{\cos nx}{n^2} \bigg|_0^{\pi} \right] = \frac{2}{\pi} \left[\frac{(-1)^n - 1}{n^2} \right]$$

$$= \begin{cases} -\frac{4}{\pi n^2}, & n \text{ is odd,} \\ 0, & n \text{ is even.} \end{cases}$$

For $n \geq 1$ it may be checked that $b_n = 0$ (see also Section 2.2). This being so, the Fourier series of $f(x) = |x|$ is given by

$$|x| \sim \frac{\pi}{2} - \sum_{k=1}^{\infty} \frac{4}{\pi(2k-1)^2} \cos(2k-1)x.$$

One of the basic properties of orthonormal series is that of the nearest vector in a subspace spanned by the partial sums of the series (see Theorem 1.17). A simple translation of Theorem 1.17 to Fourier series gives us this next result.

Theorem 2.3: *Let $f \in E$. Then for each $m \in \mathbb{N}$ the minimum of the expression*

$$\frac{1}{\pi} \int_{-\pi}^{\pi} \left| f(x) - \left[\frac{A_0}{2} + \sum_{n=1}^{m} (A_n \cos nx + B_n \sin nx) \right] \right|^2 dx,$$

as we vary over all A_n and B_n (in $\mathbb{C}$) is uniquely attained by $A_n = a_n$ and $B_n = b_n$, $n \leq m$, where the a_n and b_n are the Fourier coefficients of f.

We will later use this fact.

Exercises

1. Find the Fourier series of each of the following functions.

   (a) $f(x) = |\sin x|$ (b) $f(x) = \begin{cases} 0, & -\pi \leq x \leq 0 \\ e^x, & 0 < x \leq \pi \end{cases}$

(c) $f(x) = \begin{cases} \cos x, & -\pi < x < 0 \\ \sin x, & 0 < x < \pi \end{cases}$ (d) $f(x) = \begin{cases} 0, & -\pi \leq x \leq 0 \\ x, & 0 < x \leq \pi \end{cases}$

2. For each real p, $-\pi \leq p \leq \pi$, find the Fourier series of the function
$$f_p(x) = \begin{cases} 0, & -\pi \leq x \leq p, \\ 1, & p < x \leq \pi. \end{cases}$$

3. Let f be a 2π-periodic piecewise continuous function and let
$$f(x) \sim \frac{a_0}{2} + \sum_{n=1}^{\infty} [a_n \cos nx + b_n \sin nx]$$
denote its Fourier series.

   (a) Set $g(x) = f(x + \pi)$ for all $x \in \mathbb{R}$, and let
$$g(x) \sim \frac{A_0}{2} + \sum_{n=1}^{\infty} [A_n \cos nx + B_n \sin nx]$$
denote the Fourier series of g. Express A_n and B_n in terms of a_n and b_n.

   (b) Define $h(x) = f(x) \cos x$, and let
$$h(x) \sim \frac{\alpha_0}{2} + \sum_{n=1}^{\infty} [\alpha_n \cos nx + \beta_n \sin nx]$$
denote the Fourier series of h. Express α_n and β_n in terms of a_n and b_n.

4. Let θ be a given angle. Prove that the sequence
$$\left\{ \frac{1}{\sqrt{2}}, \cos(x+\theta), \sin(x+\theta), \cos(2x+\theta), \sin(2x+\theta), \ldots \right\}$$
is an orthonormal system in the space E equipped with the inner product
$$\langle f, g \rangle = \frac{1}{\pi} \int_{-\pi}^{\pi} f(x) \overline{g(x)} \, dx.$$

5. Let $f \in E$ and
$$f(x) \sim \frac{a_0}{2} + \sum_{n=1}^{\infty} [a_n \cos nx + b_n \sin nx]$$
denote the Fourier series of f. Prove that there exist $\{A_n\}_{n=0}^{\infty}$ and $\{\alpha_n\}_{n=0}^{\infty}$, where $-\frac{\pi}{2} < \alpha_n \leq \frac{\pi}{2}$, such that
$$f(x) \sim \frac{a_0}{2} + \sum_{n=1}^{\infty} [a_n \cos nx + b_n \sin nx] = A_0 + \sum_{n=1}^{\infty} A_n \cos(nx - \alpha_n).$$

In a similar way, prove that there exist $\{B_n\}_{n=0}^{\infty}$ and $\{\beta_n\}_{n=1}^{\infty}$, where $-\frac{\pi}{2} < \beta_n \leq \frac{\pi}{2}$, such that

$$f(x) \sim \frac{a_0}{2} + \sum_{n=1}^{\infty}[a_n \cos nx + b_n \sin nx] = B_0 + \sum_{n=1}^{\infty} B_n \sin(nx + \beta_n).$$

2. Evenness, Oddness, and Additional Examples

The determination of the Fourier series of a given function is simpler if the function is either even or odd. We recall that a function f is said to be *even about the point* a if for each (appropriate) x we have

$$f(a - x) = f(a + x).$$

Similarly, a function f is said to be *odd about the point* a if for each x we have

$$f(a - x) = -f(a + x).$$

If we do not explicitly specify the point a, then we almost always mean that $a = 0$, and it should be so understood. In this case f is even if for all x, $f(-x) = f(x)$, and f is odd if for all x, $f(-x) = -f(x)$. Examples of some standard even functions are

$$f(x) = x^{2n}, \quad f(x) = \cos nx, \quad n \in \mathbb{N},$$

and some standard odd functions are

$$f(x) = x^{2n-1}, \quad f(x) = \sin nx, \quad n \in \mathbb{N}.$$

(From these examples we see the source of the names "even" and "odd".) The following properties of even and odd functions are easily seen to hold.

(a) The product of two even functions is an even function.

(b) The product of two odd functions is an even function.

(c) The product of an even function and an odd function is an odd function.

(d) If g is an odd function, then for any $b > 0$,

$$\int_{-b}^{b} g(x)dx = 0.$$

(e) If g is an even function, then for any $b > 0$,

$$\int_{-b}^{b} g(x)dx = 2\int_{0}^{b} g(x)dx.$$

From these properties we obtain the following results which show that in case of evenness or oddness half the Fourier coefficients vanish.

Proposition 2.4: *Assume $f \in E$.*

(a) *If f is even then the Fourier series of f has the form*

$$f(x) \sim \frac{a_0}{2} + \sum_{n=1}^{\infty} a_n \cos nx$$

where

$$a_n = \frac{2}{\pi} \int_0^{\pi} f(x) \cos nx \, dx.$$

Such a series is called a cosine series.

(b) *If f is odd then the Fourier series of f has the form*

$$f(x) \sim \sum_{n=1}^{\infty} b_n \sin nx$$

where

$$b_n = \frac{2}{\pi} \int_0^{\pi} f(x) \sin nx \, dx.$$

Such a series is called a sine series.

Example 2.4: Let $f(x) = \text{sgn}(x)$ denote the "sign" function on $[-\pi, \pi]$. That is,

$$f(x) = \begin{cases} 1, & 0 < x \leq \pi, \\ 0, & x = 0, \\ -1, & -\pi \leq x < 0. \end{cases}$$

Obviously f is odd and thus $a_n = 0$ for all n. The coefficients b_n must be calculated.

$$b_n = \frac{1}{\pi} \int_{-\pi}^{\pi} \text{sgn}(x) \sin nx \, dx = \frac{2}{\pi} \int_0^{\pi} \sin nx \, dx$$

$$= \frac{2}{\pi} \left[-\frac{\cos nx}{n} \Big|_0^{\pi} \right] = \frac{2}{\pi} \left[\frac{1 - (-1)^n}{n} \right]$$

$$= \begin{cases} 0, & n \text{ is even}, \\ \frac{4}{n\pi}, & n \text{ is odd}, \end{cases}$$

and thus

$$f(x) \sim \sum_{k=1}^{\infty} \frac{4}{(2k-1)\pi} \sin(2k-1)x.$$

Example 2.5: Let $f(x) = x^2$ on the interval $[-\pi, \pi]$. The function f is even and therefore $b_n = 0$ for all n. For the calculation of $a_n, n \geq 1$, we use integration by parts.

$$a_n = \frac{1}{\pi} \int_{-\pi}^{\pi} x^2 \cos nx \, dx = \frac{2}{\pi} \int_0^{\pi} x^2 \cos nx \, dx$$

$$= \frac{2}{\pi} \left[\frac{x^2 \sin nx}{n} \bigg|_0^{\pi} - \frac{1}{n} \int_0^{\pi} 2x \sin nx \, dx \right] = -\frac{4}{\pi n} \int_0^{\pi} x \sin nx \, dx$$

$$= -\frac{4}{\pi n} \left[-\frac{x \cos nx}{n} \bigg|_0^{\pi} + \frac{1}{n} \int_0^{\pi} \cos nx \, dx \right]$$

$$= -\frac{4}{\pi n} \left[-\frac{\pi (-1)^n}{n} + \frac{\sin nx}{n^2} \bigg|_0^{\pi} \right] = \frac{4(-1)^n}{n^2}.$$

We also calculate a_0.

$$a_0 = \frac{1}{\pi} \int_{-\pi}^{\pi} x^2 \, dx = \frac{x^3}{3\pi} \bigg|_{-\pi}^{\pi} = \frac{2\pi^2}{3}.$$

Therefore

$$x^2 \sim \frac{\pi^2}{3} + \sum_{n=1}^{\infty} \frac{4(-1)^n}{n^2} \cos nx.$$

Exercises

1. Find the Fourier series of each of the following functions.
   (a) $f(x) = x|x|$ (b) $f(x) = \pi^2 - x^2$ (c) $f(x) = \sin \frac{x}{2}$
2. Find the Fourier series of $f_p(x) = \cos px$, for $0 \leq p \leq \pi$.
3. Let $f \in E$ and

$$f(x) \sim \frac{a_0}{2} + \sum_{n=1}^{\infty} [a_n \cos nx + b_n \sin nx]$$

denote its Fourier series. Define the two functions

$$g(x) = \frac{f(x) + f(-x)}{2}, \qquad h(x) = \frac{f(x) - f(-x)}{2}.$$

Find the Fourier series of g and of h.

3. Complex Fourier Series

Till now we have dealt with the orthonormal system composed of the real functions $\sin nx$ and $\cos nx$. We now present a very important orthonormal system whose elements are complex-valued functions. The appropriate associated inner product is slightly different and is given by

$$\langle f, g \rangle = \frac{1}{2\pi} \int_{-\pi}^{\pi} f(x)\overline{g(x)}\, dx.$$

We recall Euler's formula which states that for every $x \in \mathbb{R}$,

$$e^{ix} = \cos x + i \sin x.$$

It is easy to verify that the set of functions $\{e^{inx}\}_{n=-\infty}^{\infty}$ form an orthonormal system with respect to the above inner product. Thus for each $f \in E$, the appropriate series associated with this orthonormal system is

(2.4) $$f(x) \sim \sum_{n=-\infty}^{\infty} c_n e^{inx}$$

where

(2.5) $$c_n = \frac{1}{2\pi} \int_{-\pi}^{\pi} f(x) e^{-inx}\, dx, \qquad n = 0, \pm 1, \pm 2, \ldots.$$

Definition 2.5: *Let $f \in E$. Then the series (2.4) is called the* **complex** Fourier series *of f where the coefficients c_n are defined as in (2.5).*

The complex Fourier series and the usual Fourier series are in fact one and the same. That is, they are different ways of writing the same series. To see this, note that

$$e^{i0x} = 1,$$
$$e^{inx} = \cos nx + i \sin nx, \quad n = 1, 2, 3, \ldots,$$
$$e^{-inx} = \cos nx - i \sin nx, \quad n = 1, 2, 3, \ldots.$$

Let a_n and b_n be as in (2.3). Then for every $n = 1, 2, 3, \ldots$,

$$c_n = \frac{1}{2\pi} \int_{-\pi}^{\pi} f(x) e^{-inx}\, dx$$
$$= \frac{1}{2\pi} \left[\int_{-\pi}^{\pi} f(x) \cos nx\, dx - i \int_{-\pi}^{\pi} f(x) \sin nx\, dx \right]$$
$$= \frac{a_n - i b_n}{2}$$

and
$$c_{-n} = \frac{1}{2\pi} \int_{-\pi}^{\pi} f(x) e^{inx} dx$$
$$= \frac{1}{2\pi} \left[\int_{-\pi}^{\pi} f(x) \cos nx\, dx + i \int_{-\pi}^{\pi} f(x) \sin nx\, dx \right]$$
$$= \frac{a_n + ib_n}{2},$$

from which it follows that
$$a_n = c_n + c_{-n}, \qquad b_n = i(c_n - c_{-n}).$$

For $n = 0$,
$$c_0 = \frac{1}{2\pi} \int_{-\pi}^{\pi} f(x) e^{-i0x}\, dx = \frac{1}{2\pi} \int_{-\pi}^{\pi} f(x)\, dx = \frac{a_0}{2}.$$

Thus
$$f(x) \sim \sum_{n=-\infty}^{\infty} c_n e^{inx} = c_0 + \sum_{n=1}^{\infty} \left[c_n e^{inx} + c_{-n} e^{-inx} \right]$$
$$= c_0 + \sum_{n=1}^{\infty} \left[c_n (\cos nx + i \sin nx) + c_{-n} (\cos nx - i \sin nx) \right]$$
$$= c_0 + \sum_{n=1}^{\infty} \left[(c_n + c_{-n}) \cos nx + i(c_n - c_{-n}) \sin nx \right]$$
$$= \frac{a_0}{2} + \sum_{n=1}^{\infty} \left[a_n \cos nx + b_n \sin nx \right].$$

So we see that there is no real difference between the complex Fourier series and the usual Fourier series. If so, why introduce the former? Firstly, both are used and the reader should be aware of both forms. Secondly, the complex Fourier series is sometimes more useful, and sometimes more natural. It emphasizes the fact that we can (and often should) view a 2π-periodic function as if it were a function defined on the boundary of the unit circle in the complex plane.

Example 2.6: Let us calculate the complex Fourier series of $f(x) = x$. We will use integration by parts. For each $n \in \mathbb{Z}$, $n \neq 0$,
$$c_n = \frac{1}{2\pi} \int_{-\pi}^{\pi} x e^{-inx} dx = \frac{1}{2\pi} \left[\frac{x e^{-inx}}{-in} \Big|_{-\pi}^{\pi} + \frac{1}{in} \int_{-\pi}^{\pi} e^{-inx} dx \right]$$
$$= \frac{1}{2\pi} \left[\frac{\pi e^{-in\pi}}{-in} - \frac{-\pi e^{in\pi}}{-in} + \frac{e^{-inx}}{n^2} \Big|_{-\pi}^{\pi} \right] = \frac{1}{2\pi} \left[\frac{\pi e^{-in\pi} + \pi e^{in\pi}}{-in} \right]$$
$$= \frac{i \cos n\pi}{n} = \frac{(-1)^n i}{n}.$$

For $n = 0$,
$$c_0 = \frac{1}{2\pi} \int_{-\pi}^{\pi} x \, dx = 0.$$

Thus
$$x \sim \sum_{n=1}^{\infty} \frac{(-1)^n i}{n} e^{inx} + \sum_{n=-1}^{-\infty} \frac{(-1)^n i}{n} e^{inx}.$$

Please note that we could also have obtained this result from the standard Fourier series of $f(x) = x$ (which we found in Example 2.1) and the equalities $c_n = \frac{a_n - ib_n}{2}$ and $c_{-n} = \frac{a_n + ib_n}{2}$, $n \in \mathbb{N}$ (as proven above).

Exercises

1. Let $f \in E$. Find the complex Fourier series of $\mathrm{Re}(f)$ on the basis of the complex Fourier series of f.

2. Let
$$f(x) = \begin{cases} 0, & -\pi \leq x < 0, \\ e^{ix}, & 0 \leq x < \pi. \end{cases}$$
Find the complex Fourier series of f.

3. Let $f \in E$ and
$$f(x) \sim \sum_{n=-\infty}^{\infty} c_n e^{inx}$$
be the complex Fourier series of f. Determine the complex Fourier series of $f(\overline{x})$, $\overline{f(x)}$, and $f(-x)$.

4. Let $f, g \in E$ be 2π-periodic functions, and
$$f(x) \sim \sum_{n=-\infty}^{\infty} a_n e^{inx}, \qquad g(x) \sim \sum_{n=-\infty}^{\infty} b_n e^{inx}$$
be the complex Fourier series of f and g. For each $x \in \mathbb{R}$ we define
$$h(x) = \frac{1}{2\pi} \int_{-\pi}^{\pi} f(x-t)g(t) \, dt.$$

   (a) Prove that h is piecewise continuous and 2π-periodic.
   (b) Let $h(x) \sim \sum_{n=-\infty}^{\infty} c_n e^{inx}$ be the complex Fourier series of h. Prove that $c_n = a_n b_n$ for all $n \in \mathbb{Z}$.

5. Find the complex Fourier series of $f(x) = e^x$.

6. Let $f(x) = \sum_{n=-\infty}^{\infty} c_n e^{inx}$. Prove the following claims:
  (a) If f real-valued, then $c_{-n} = \overline{c_n}$.
  (b) If f is purely imaginary, then $c_{-n} = -\overline{c_n}$.
  (c) If f is real-valued and even, then the c_n are real.
  (d) If f is real-valued and odd, then the c_n are purely imaginary.

4. Pointwise Convergence and Dirichlet's Theorem

The closure property of the trigonometric orthonormal system (a fact still to be proved) implies that the Fourier series of each $f \in E$ converges in norm to f. In other words, if the a_n and b_n are the Fourier coefficients of f, then

$$\lim_{m \to \infty} \left\| f(x) - \left(\frac{a_0}{2} + \sum_{n=1}^{m} [a_n \cos nx + b_n \sin nx] \right) \right\| = 0.$$

We can also write this more explicitly in the form

$$\lim_{m \to \infty} \int_{-\pi}^{\pi} \left| f(x) - \left(\frac{a_0}{2} + \sum_{n=1}^{m} [a_n \cos nx + b_n \sin nx] \right) \right|^2 dx = 0.$$

These formulae strengthen the notion that the partial sums of the Fourier series are better and better approximations of the function f, in some sense. However, as we already saw in Example 2.1, this does not necessarily imply convergence at every point. In this section we present conditions which guarantee the important property of pointwise convergence of the Fourier series of f to f. That is to say, conditions under which we have

$$f(x) = \frac{a_0}{2} + \sum_{n=1}^{\infty} [a_n \cos nx + b_n \sin nx]$$

for each $x \in [-\pi, \pi]$. In order to obtain this type of convergence, it is necessary to limit ourselves to a more restricted set of functions. Even in this case the pointwise convergence will not hold at every point of $[-\pi, \pi]$ (but only at "good" points of the interval).

We define E' to be the space of functions $f : [-\pi, \pi] \to \mathbb{C}$ which satisfy the following conditions:

1. $f \in E$.
2. At each $x \in [-\pi, \pi)$, the following limit exists (and is finite):

$$\lim_{h \to 0^+} \frac{f(x+h) - f(x+)}{h}.$$

3. At each $x \in (-\pi, \pi]$, the following limit exists (and is finite):
$$\lim_{h \to 0^+} \frac{f(x-) - f(x-h)}{h}.$$

The expressions $f(x-)$ and $f(x+)$ represent the left and right limits of f at x, respectively. They exist since $f \in E$. The latter two conditions are simply the appropriate one-sided derivatives of f at each point in $[-\pi, \pi]$.

Theorem 2.6: (Dirichlet's Theorem) *Let $f \in E'$. Then for each $x \in (-\pi, \pi)$ the Fourier series of f converges to the value*
$$\frac{f(x-) + f(x+)}{2}.$$
At both endpoints $x = \pm \pi$ the series converges to
$$\frac{f(\pi-) + f((-\pi)+)}{2}.$$

Remarks: (a) The endpoints $x = \pm\pi$ are not really special cases. If, as we mentioned earlier, we assume that f is defined on all of $\mathbb{R}$ and is 2π-periodic, then from the periodicity we have $f(\pi+) = f((-\pi)+)$, and thus
$$\frac{f(\pi-) + f(\pi+)}{2} = \frac{f(\pi-) + f((-\pi)+)}{2}.$$
A similar situation holds at the other endpoint $x = -\pi$.

(b) If f is continuous at a point x then $f(x-) = f(x+)$, and thus $\frac{f(x-)+f(x+)}{2} = f(x)$. The Fourier series of f therefore converges to $f(x)$ at this point. It thus follows that if, in addition to the conditions of the theorem, f is continuous on $[-\pi, \pi]$ and satisfies $f(-\pi) = f(\pi)$, then the Fourier series of f converges to $f(x)$ at each and every $x \in [-\pi, \pi]$.

The proof of Dirichlet's Theorem is long and complicated. This being so, we divide it into parts and first prove a number of claims which will hopefully allow for a more convenient presentation. Some of these claims are important in and of themselves and will be used later for other purposes.

Without loss of generality we assume that $f \in E'$ is defined on all of $\mathbb{R}$, and is 2π-periodic. For each natural number m, we denote the partial sum of the Fourier series of f by
$$S_m(x) = \frac{a_0}{2} + \sum_{n=1}^{m} [a_n \cos nx + b_n \sin nx].$$

Proposition 2.7: $\quad S_m(x) = \dfrac{1}{\pi} \displaystyle\int_{-\pi}^{\pi} f(x+t) \left[\dfrac{1}{2} + \sum_{n=1}^{m} \cos nt \right] dt.$

Proof: From the definition of the coefficients a_n and b_n we have

$$S_m(x) = \frac{a_0}{2} + \sum_{n=1}^{m} [a_n \cos nx + b_n \sin nx]$$

$$= \frac{1}{2\pi} \int_{-\pi}^{\pi} f(s) \, ds + \sum_{n=1}^{m} \left[\frac{1}{\pi} \int_{-\pi}^{\pi} f(s) \cos ns \, ds \cdot \cos nx \right.$$

$$\left. + \frac{1}{\pi} \int_{-\pi}^{\pi} f(s) \sin ns \, ds \cdot \sin nx \right]$$

$$= \frac{1}{\pi} \int_{-\pi}^{\pi} f(s) \left[\frac{1}{2} + \sum_{n=1}^{m} [\cos ns \cos nx + \sin ns \sin nx] \right] ds$$

$$= \frac{1}{\pi} \int_{-\pi}^{\pi} f(s) \left[\frac{1}{2} + \sum_{n=1}^{m} \cos n(s-x) \right] ds.$$

Substituting $t = s - x$ gives us

$$S_m(x) = \frac{1}{\pi} \int_{-\pi-x}^{\pi-x} f(x+t) \left[\frac{1}{2} + \sum_{n=1}^{m} \cos nt \right] dt.$$

For every 2π-periodic function g, and any real a

$$\int_{-\pi+a}^{\pi+a} g(t) \, dt = \int_{-\pi}^{\pi} g(t) \, dt.$$

(The reader should verify this fact.) Thus

$$S_m(x) = \frac{1}{\pi} \int_{-\pi}^{\pi} f(x+t) \left[\frac{1}{2} + \sum_{n=1}^{m} \cos nt \right] dt.$$

∎

Proposition 2.8: *For each natural number m we have*

$$\frac{1}{2} + \cos t + \cos 2t + \cdots + \cos mt = \frac{\sin(m+\frac{1}{2})t}{2 \sin \frac{1}{2} t}.$$

Proof: We use the trigonometric identity

$$\cos \alpha \sin \beta = \tfrac{1}{2}[\sin(\alpha + \beta) - \sin(\alpha - \beta)],$$

from which it immediately follows that for every natural number k and real t

$$\cos kt \sin \tfrac{1}{2} t = \tfrac{1}{2}[\sin(k + \tfrac{1}{2})t - \sin(k - \tfrac{1}{2})t].$$

Thus

$$\sin \tfrac{1}{2}t \left[\tfrac{1}{2} + \cos t + \cos 2t + \cdots + \cos mt\right] = \tfrac{1}{2}\left[\sin \tfrac{1}{2}t + \left(\sin \tfrac{3}{2}t - \sin \tfrac{1}{2}t\right)\right.$$
$$\left. + \left(\sin \tfrac{5}{2}t - \sin \tfrac{3}{2}t\right) + \cdots + \left(\sin(m+\tfrac{1}{2})t - \sin(m-\tfrac{1}{2})t\right)\right]$$
$$= \tfrac{1}{2}\sin(m+\tfrac{1}{2})t.$$

Dividing both sides by $\sin \tfrac{1}{2}t$ gives us the desired conclusion. ∎

The function $D_m(t) = \tfrac{1}{2} + \sum_{n=1}^{m} \cos nt$ is called the *Dirichlet kernel* (see Figure 2.1). From Proposition 2.8 we see that $D_m(t) = \frac{\sin(m+\tfrac{1}{2})t}{2\sin \tfrac{1}{2}t}$. The points $t = 2n\pi$, where the denominator vanishes, are removable points of discontinuity.

Proposition 2.9: $\int_0^\pi D_m(t)\,dt = \dfrac{\pi}{2}.$

Proof:
$$\int_0^\pi D_m(t)\,dt = \int_0^\pi [\tfrac{1}{2} + \cos t + \cos 2t + \cdots + \cos mt]\,dt.$$
For every positive integer $n = 1, 2, 3, \ldots$, $\int_0^\pi \cos nt\,dt = 0$. Therefore
$$\int_0^\pi D_m(t)\,dt = \int_0^\pi \tfrac{1}{2}\,dt = \dfrac{\pi}{2}. \qquad \blacksquare$$

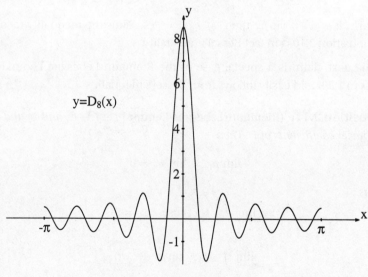

Fig. 2.1

The next proposition is a special case of Bessel's inequality which we proved in Theorem 1.19.

Proposition 2.10: (**Bessel's Inequality**) *Let $f \in E$ and a_n, b_n be the Fourier coefficients of f. Then*

$$\frac{|a_0|^2}{2} + \sum_{n=1}^{\infty} \left(|a_n|^2 + |b_n|^2\right) \leq \|f\|^2.$$

Proof: We prove this inequality by explaining how it is a special case of Bessel's inequality from Section 1.5. The general form of Bessel's inequality used there is

$$\sum_{n=1}^{\infty} |\langle f, e_n \rangle|^2 \leq \|f\|^2$$

where $\{e_n\}_{n=1}^{\infty}$ is an orthonormal system. For $f \in E$,

$$\|f\|^2 = \frac{1}{\pi} \int_{-\pi}^{\pi} |f(x)|^2 dx.$$

With $e_n(x) = \frac{1}{\sqrt{2}}$ we have

$$|\langle f, e_n \rangle|^2 = \left| \frac{1}{\pi} \int_{-\pi}^{\pi} f(x) \frac{1}{\sqrt{2}} dx \right|^2 = \left| \frac{a_0}{\sqrt{2}} \right|^2 = \frac{|a_0|^2}{2}.$$

If $e_n(x) = \cos nx$, $n \geq 1$, then

$$|\langle f, e_n \rangle|^2 = \left| \frac{1}{\pi} \int_{-\pi}^{\pi} f(x) \cos nx \, dx \right|^2 = |a_n|^2.$$

Similarly, if $e_n(x) = \sin nx$ then $|\langle f, e_n \rangle| = |b_n|$. Thus the inequality recorded as Proposition 2.10 is in fact Bessel's inequality. ∎

This next claim is a special case of the Riemann-Lebesgue Lemma (see Theorem 1.20) and easily follows from Bessel's inequality.

Proposition 2.11: (**Riemann-Lebesgue Lemma**) *Let $f \in E$, and a_n and b_n be the Fourier coefficients of f. Then*

$$\lim_{n \to \infty} a_n = \lim_{n \to \infty} b_n = 0.$$

That is

$$\lim_{n \to \infty} \int_{-\pi}^{\pi} f(x) \cos nx \, dx = 0,$$

$$\lim_{n \to \infty} \int_{-\pi}^{\pi} f(x) \sin nx \, dx = 0.$$

Proposition 2.12: *For each piecewise continuous function g defined on $[0, \pi]$, we have*

$$\lim_{m \to \infty} \int_0^{\pi} g(t) \sin(m + \tfrac{1}{2}) t \, dt = 0.$$

Chapter 2: Fourier Series

Proof: We define the two functions

$$h_1(t) = \begin{cases} g(t) \cos \frac{t}{2}, & 0 \leq t \leq \pi, \\ 0, & -\pi \leq t < 0. \end{cases}$$

and

$$h_2(t) = \begin{cases} g(t) \sin \frac{t}{2}, & 0 \leq t \leq \pi, \\ 0, & -\pi \leq t < 0. \end{cases}$$

Since g is piecewise continuous on $[0, \pi]$, it follows that both h_1 and h_2 are piecewise continuous on $[-\pi, \pi]$. Now

$$\int_0^\pi g(t) \sin(m + \tfrac{1}{2})t \, dt = \int_0^\pi g(t) \cos \tfrac{t}{2} \sin mt \, dt + \int_0^\pi g(t) \sin \tfrac{t}{2} \cos mt \, dt$$

$$= \int_{-\pi}^\pi h_1(t) \sin mt \, dt + \int_{-\pi}^\pi h_2(t) \cos mt \, dt.$$

From Proposition 2.11 (Riemann-Lebesgue Lemma), each of the last two integrals tends to zero as m tends to infinity. This gives the desired result. ∎

Proof of Theorem 2.6: We wish to prove that for each x,

$$\lim_{m \to \infty} S_m(x) = \frac{f(x-) + f(x+)}{2}.$$

Choose an x and consider it as fixed throughout this proof. Define the function

$$g(t) = \frac{f(x+t) - f(x+)}{2 \sin \frac{t}{2}}.$$

Obviously, g is piecewise continuous on $(0, \pi]$. In addition

$$\lim_{t \to 0+} g(t) = \lim_{t \to 0+} \frac{f(x+t) - f(x+)}{2 \sin \frac{t}{2}}$$

$$= \lim_{t \to 0+} \left[\frac{f(x+t) - f(x+)}{t} \right] \left[\frac{t}{2 \sin \frac{t}{2}} \right].$$

The last limit exists and is finite because

$$\lim_{t \to 0+} \frac{t}{2 \sin \frac{t}{2}} = 1$$

and the limit

$$\lim_{t \to 0+} \frac{f(x+t) - f(x+)}{t}$$

exists since $f \in E'$. Thus, as a consequence of the assumption of the theorem, g is piecewise continuous on all of $[0, \pi]$. From Proposition 2.12

$$\lim_{m \to \infty} \frac{1}{\pi} \int_0^\pi g(t) \sin(m + \tfrac{1}{2})t \, dt = 0.$$

From the definition of g we have

$$\lim_{m \to \infty} \left[\frac{1}{\pi} \int_0^\pi \frac{f(x+t)}{2 \sin \frac{t}{2}} \sin(m + \tfrac{1}{2})t \, dt - \frac{1}{\pi} \int_0^\pi \frac{f(x+)}{2 \sin \frac{t}{2}} \sin(m + \tfrac{1}{2})t \, dt \right] = 0.$$

Furthermore, from Proposition 2.9

$$\frac{1}{\pi} \int_0^\pi \frac{f(x+)}{2 \sin \frac{t}{2}} \sin(m + \tfrac{1}{2})t \, dt = f(x+) \frac{1}{\pi} \int_0^\pi \frac{\sin(m + \tfrac{1}{2})t}{2 \sin \frac{t}{2}} \, dt$$

$$= f(x+) \frac{1}{\pi} \int_0^\pi D_m(t) \, dt$$

$$= \frac{f(x+)}{2}.$$

Thus
$$\lim_{m \to \infty} \frac{1}{\pi} \int_0^\pi \frac{f(x+t)}{2 \sin \frac{t}{2}} \sin(m + \tfrac{1}{2})t \, dt = \frac{f(x+)}{2}.$$

By analogous reasoning we also show that

$$\lim_{m \to \infty} \frac{1}{\pi} \int_{-\pi}^0 \frac{f(x+t)}{2 \sin \frac{t}{2}} \sin(m + \tfrac{1}{2})t \, dt = \frac{f(x-)}{2}.$$

Thus, using Propositions 2.7 and 2.8,

$$\lim_{m \to \infty} S_m(x) = \lim_{m \to \infty} \frac{1}{\pi} \int_{-\pi}^\pi f(x+t) D_m(t) \, dt$$

$$= \lim_{m \to \infty} \frac{1}{\pi} \int_{-\pi}^0 f(x+t) D_m(t) \, dt + \lim_{m \to \infty} \frac{1}{\pi} \int_0^\pi f(x+t) D_m(t) \, dt$$

$$= \frac{f(x-) + f(x+)}{2}. \qquad \blacksquare$$

We now know that if a function satisfies the assumptions of Dirichlet's Theorem then its Fourier series converges at every point of $\mathbb{R}$. In addition, the series converges to the value of the function at that point, if the function is continuous there. At points of discontinuity the series converges to the best value which one could expect, namely the average of the one-sided limits of the function at the point.

Chapter 2: Fourier Series

Example 2.7: Let $f(x) = x$ on the interval $[-\pi, \pi]$. In Example 2.1 we found the Fourier series of f to be

$$f(x) \sim \sum_{n=1}^{\infty} \frac{2(-1)^{n+1}}{n} \sin nx.$$

By Dirichlet's Theorem this series converges to x at each $x \in (-\pi, \pi)$, and converges to zero at the endpoints $x = \pm \pi$.

If we substitute $x = \frac{\pi}{2}$ in the series we obtain

$$\frac{\pi}{2} = \sum_{n=1}^{\infty} \frac{2(-1)^{n+1}}{n} \sin n\frac{\pi}{2}$$

$$= 2 \left[\frac{\sin \frac{\pi}{2}}{1} - \frac{\sin \frac{2\pi}{2}}{2} + \frac{\sin \frac{3\pi}{2}}{3} - \frac{\sin \frac{4\pi}{2}}{4} + \cdots \right]$$

$$= 2 \left[1 - \frac{1}{3} + \frac{1}{5} - \frac{1}{7} + \cdots \right]$$

and therefore

$$\frac{\pi}{4} = 1 - \frac{1}{3} + \frac{1}{5} - \frac{1}{7} + \cdots .$$

By substituting $x = \frac{\pi}{4}$ and similar calculations we obtain

$$\frac{\pi}{2\sqrt{2}} = 1 + \frac{1}{3} - \frac{1}{5} - \frac{1}{7} + \frac{1}{9} + \frac{1}{11} - \frac{1}{13} - \frac{1}{15} + \cdots .$$

(We can also obtain this formula from the Fourier series of $\text{sgn}(x)$ (see Example 2.4) by substituting the value $x = \frac{\pi}{2}$.)

Example 2.8: In Example 2.5 we determined the Fourier series of $f(x) = x^2$:

$$x^2 \sim \frac{\pi^2}{3} + \sum_{n=1}^{\infty} \frac{4(-1)^n}{n^2} \cos nx.$$

The function $f(x) = x^2$ satisfies the condition of Dirichlet's Theorem on $[-\pi, \pi]$. In addition, the 2π-periodic continuation of f is continuous on all of $\mathbb{R}$, and we can therefore write

(2.6) $$x^2 = \frac{\pi^2}{3} + \sum_{n=1}^{\infty} \frac{4(-1)^n}{n^2} \cos nx$$

for $x \in [-\pi, \pi]$.

If we substitute $x = 0$ in this formula we obtain

$$0 = \frac{\pi^2}{3} + \sum_{n=1}^{\infty} \frac{4(-1)^n}{n^2}$$

and therefore

$$(2.7) \qquad \sum_{n=1}^{\infty} \frac{(-1)^{n+1}}{n^2} = \frac{\pi^2}{12}.$$

If we substitute $x = \pi$ in formula (2.6) we obtain

$$\sum_{n=1}^{\infty} \frac{1}{n^2} = \frac{\pi^2}{6}.$$

The last identity can be obtained from (2.7) (and vice versa), as follows:

$$S = \sum_{n=1}^{\infty} \frac{1}{n^2} = \sum_{n=1}^{\infty} \frac{(-1)^{n+1}}{n^2} + 2\sum_{k=1}^{\infty} \frac{1}{(2k)^2} = \frac{\pi^2}{12} + \frac{1}{2}\sum_{k=1}^{\infty} \frac{1}{k^2} = \frac{\pi^2}{12} + \frac{1}{2}S.$$

Hence $S = \frac{\pi^2}{12} + \frac{1}{2}S$, and thus $S = \frac{\pi^2}{6}$.

Exercises

1. Set $f(x) = 1 - x^2$ in the interval $[-\pi, \pi]$ and let

    $$f(x) \sim \frac{a_0}{2} + \sum_{n=1}^{\infty} [a_n \cos nx + b_n \sin nx]$$

    be the Fourier series of f.
    (a) Calculate the a_n and b_n.
    (b) To what values does the Fourier series of f converge at the points $x = 5\pi$ and $x = 6\pi$? Explain.

2. For each real number $p \neq 0$, set $f_p(x) = e^{px}$ in the interval $[-\pi, \pi]$. Let

    $$f_p(x) \sim \frac{a_0}{2} + \sum_{n=1}^{\infty} [a_n \cos nx + b_n \sin nx]$$

    denote the Fourier series of f_p.
    (a) Calculate a_n and b_n.
    (b) Determine $\sum_{n=0}^{\infty} a_n$ and $\sum_{n=0}^{\infty} (-1)^n a_n$.

3. Assume f satisfies the assumptions of Dirichlet's Theorem. Determine the following limits:
    (a) $\lim_{n \to \infty} \frac{1}{\pi} \int_{-\pi}^{\pi} f(t) \sin nt \, dt$
    (b) $\lim_{n \to \infty} \frac{1}{\pi} \int_{-\pi}^{\pi} f(t) \sin(n - \frac{1}{2})t \, dt$
    (c) $\lim_{n \to \infty} \frac{1}{\pi} \int_{-\pi}^{\pi} \frac{f(t)}{t} \sin(n - \frac{1}{2})t \, dt$

4. Find the Fourier series of

$$f(x) = \begin{cases} x - [x], & x \text{ is not an integer,} \\ \frac{1}{2}, & x \text{ is an integer.} \end{cases}$$

To what values does the Fourier series converge at the points $x = 5$, $x = 3$, and $x = 1.5$?

5. For each natural number m, let $D_m(t) = \frac{1}{2} + \sum_{n=1}^{m} \cos nt$.
   (a) Calculate $\int_{-\pi}^{\pi} D_m(t) \sin 100t\, dt$.
   (b) Determine $\frac{1}{\pi} \int_{-\pi}^{\pi} [D_m(t)]^2\, dt$, for $m = 100$.
   (c) Let $g(t) = \begin{cases} \frac{\sin \frac{1}{2}t}{t}, & t \neq 0, \\ \frac{1}{2}, & t = 0. \end{cases}$

   Calculate $\lim_{m \to \infty} \frac{1}{\pi} \int_{-\pi}^{\pi} D_m(t) g(t)\, dt$.

6. Let $f(x) = xe^{ix}$. Find the complex Fourier series of f and calculate $F(\pi)$, $F(-\pi)$, and $F(\frac{\pi}{4})$, where

$$F(x) = \sum_{n=-\infty}^{\infty} c_n e^{inx}$$

is the complex Fourier series of f.

7. Let

$$f(x) = \begin{cases} 2 + \frac{2x}{\pi}, & -\pi < x < 0, \\ 2, & 0 < x < \pi. \end{cases}$$

   (a) Calculate the coefficients a_n and b_n of the Fourier series f.
   (b) Let

$$g(x) = \frac{a_0}{2} + \sum_{n=1}^{\infty} [a_n \cos nx + b_n \sin nx], \quad -\infty < x < \infty.$$

   Sketch the graph of g on the interval $[-3\pi, 3\pi]$.

8. Prove that if g is a piecewise continuous 2π-periodic function on $\mathbb{R}$, then for every real a

$$\int_{-\pi+a}^{\pi+a} g(t)\, dt = \int_{-\pi}^{\pi} g(t)\, dt.$$

5. Uniform Convergence

Assume $f \in E'$ and let $\{a_n\}_{n=0}^{\infty}$ and $\{b_n\}_{n=1}^{\infty}$ denote the Fourier coefficients of f. By Dirichlet's Theorem for each x

$$\lim_{m \to \infty} \left[\frac{a_0}{2} + \sum_{n=1}^{m} [a_n \cos nx + b_n \sin nx] \right] = \frac{f(x-) + f(x+)}{2}.$$

This is pointwise convergence. In the present section we consider conditions on f under which the Fourier series of f converges uniformly to f. Before doing so, we first present some formal definitions in an attempt to understand the differences between these two forms of convergence.

Definition 2.13: (Pointwise Convergence) *Let $\{f_m\}_{m=1}^{\infty}$ be a sequence of functions defined on $[a, b]$. Let f be defined on $[a, b]$. We say that the sequence $\{f_m\}_{m=1}^{\infty}$ converges pointwise to f on $[a, b]$ if for each $x \in [a, b]$ we have $\lim_{m \to \infty} f_m(x) = f(x)$. That is, for each $x \in [a, b]$ and $\epsilon > 0$ there exists a natural number $N(\epsilon, x)$ such that*

$$|f_m(x) - f(x)| < \epsilon$$

for all $m \geq N(\epsilon, x)$.

Definition 2.14: (Uniform Convergence) *Let $\{f_m\}_{m=1}^{\infty}$ be a sequence of functions defined on $[a, b]$. Let f be defined on $[a, b]$. We say that the sequence $\{f_m\}_{m=1}^{\infty}$ converges uniformly to f on $[a, b]$ if for each $\epsilon > 0$ there exists a natural number $N(\epsilon)$ such that*

$$|f_m(x) - f(x)| < \epsilon$$

for all $m \geq N(\epsilon)$, and for all $x \in [a, b]$.

These two definitions seem very similar, but there is in fact a significant difference. Uniform convergence is considerably stronger than pointwise convergence. If a sequence of functions converges uniformly then it also converges pointwise to the same function. The converse is generally not true. These two concepts of convergence describe two different ways in which the sequence $\{f_m\}_{m=1}^{\infty}$ approaches the function f. In the case of pointwise convergence one considers each $x \in [a, b]$ separately and demands for each $\epsilon > 0$ the existence of a particular $N(\epsilon, x)$ which also depends on x. It is very possible that a particular $N(\epsilon, x)$ is not suitable for other points x'. On the other hand uniform convergence is in a sense global. To each $\epsilon > 0$ there exists a number $N(\epsilon)$ which is good for all $x \in [a,b]$. Let us consider a simple example. Let $f_m(x) = x^m$ on $[0,1]$. It is easily seen that the sequence converges pointwise to the function

$$f(x) = \begin{cases} 0, & 0 \leq x < 1, \\ 1, & x = 1. \end{cases}$$

However, the sequence does not converge uniformly to f on $[0,1]$. One explanation for this fact is that for each m there exists an x_m such that $f_m(x_m) = \frac{1}{2}$. Thus if we take, for example, $\epsilon = \frac{1}{4}$ then there cannot exist an N satisfying the demands of the definition. In general, a sequence of continuous functions cannot converge uniformly to a discontinuous function (why?), but may converge pointwise to such a function.

Let us now return to our f and our particular series. The partial sums

$$S_m(x) = \frac{a_0}{2} + \sum_{n=1}^{m} [a_n \cos nx + b_n \sin nx]$$

of the Fourier series of f are a finite sum of continuous functions. Thus S_m is a continuous function for each m. In addition S_m is 2π-periodic and hence satisfies $S_m(-\pi) = S_m(\pi)$ for each m. This being so, a necessary condition on f so that the series $\{S_m\}_{m=1}^{\infty}$ converges uniformly to f on $[-\pi, \pi]$ is that f be continuous on $[-\pi, \pi]$ and also satisfy $f(-\pi) = f(\pi)$. This somewhat explains the conditions of this next theorem.

Theorem 2.15: *If f is continuous on $[-\pi, \pi]$, $f(-\pi) = f(\pi)$, and $f' \in E$, then the Fourier series of f converges uniformly to f on $[-\pi, \pi]$.*

Remark: Note that $f' \in E$ implies that $f \in E'$.

Proof: Since $f' \in E$ the function f' has a Fourier series which we write

$$f'(x) \sim \frac{\alpha_0}{2} + \sum_{n=1}^{\infty} [\alpha_n \cos nx + \beta_n \sin nx].$$

Naturally f also has a Fourier series

$$f(x) \sim \frac{a_0}{2} + \sum_{n=1}^{\infty} [a_n \cos nx + b_n \sin nx].$$

Let us now consider the relations between the α_n and β_n, and the a_n and b_n. In these calculations we use the fact that f is continuous and $f(-\pi) = f(\pi)$.

(a)
$$\alpha_0 = \frac{1}{\pi} \int_{-\pi}^{\pi} f'(x) dx = \frac{f(\pi) - f(-\pi)}{2} = 0.$$

For $n \geq 1$:

(b)
$$\alpha_n = \frac{1}{\pi}\int_{-\pi}^{\pi} f'(x)\cos nx\, dx$$
$$= \frac{1}{\pi}\left[f(x)\cos nx\Big|_{-\pi}^{\pi} + \int_{-\pi}^{\pi} f(x) n\sin nx\, dx\right]$$
$$= \frac{n}{\pi}\int_{-\pi}^{\pi} f(x)\sin nx\, dx$$
$$= nb_n.$$

(c)
$$\beta_n = \frac{1}{\pi}\int_{-\pi}^{\pi} f'(x)\sin nx\, dx$$
$$= \frac{1}{\pi}\left[f(x)\sin nx\Big|_{-\pi}^{\pi} - \int_{-\pi}^{\pi} f(x) n\cos nx\, dx\right]$$
$$= -na_n.$$

Thus we can record
$$f'(x) \sim \sum_{n=1}^{\infty}[nb_n\cos nx - na_n\sin nx].$$

Our next task is to prove that the series $\sum_{n=1}^{\infty}\sqrt{|a_n|^2 + |b_n|^2}$ converges. (This is not the series which appears in Bessel's inequality.)

$$\sum_{n=1}^{\infty}\sqrt{|a_n|^2 + |b_n|^2} = \sum_{n=1}^{\infty}\sqrt{\left|\frac{\alpha_n}{n}\right|^2 + \left|\frac{\beta_n}{n}\right|^2} = \sum_{n=1}^{\infty}\frac{1}{n}\sqrt{|\alpha_n|^2 + |\beta_n|^2}.$$

From the Cauchy-Schwarz inequality (see Example 1.10) we have

$$\sum_{n=1}^{\infty}\frac{1}{n}\sqrt{|\alpha_n|^2 + |\beta_n|^2} \le \sqrt{\sum_{n=1}^{\infty}\frac{1}{n^2}}\sqrt{\sum_{n=1}^{\infty}|\alpha_n|^2 + |\beta_n|^2}$$

and thus

$$\sum_{n=1}^{\infty}\sqrt{|a_n|^2 + |b_n|^2} \le \sqrt{\sum_{n=1}^{\infty}\frac{1}{n^2}}\sqrt{\sum_{n=1}^{\infty}|\alpha_n|^2 + |\beta_n|^2}.$$

It is well known that the series $\sum_{n=1}^{\infty}\frac{1}{n^2}$ converges, and from Bessel's inequality (Proposition 2.10) it follows that

$$\sum_{n=1}^{\infty}\left(|\alpha_n|^2 + |\beta_n|^2\right) \le \|f'\|^2 < \infty.$$

Chapter 2: Fourier Series

Thus the series $\sum_{n=1}^{\infty} \sqrt{|a_n|^2 + |b_n|^2}$ converges. From the comparison test for series it now follows that $\sum_{n=1}^{\infty} |a_n|$ and $\sum_{n=1}^{\infty} |b_n|$ converge since

$$|a_n|, |b_n| \leq \sqrt{|a_n|^2 + |b_n|^2}.$$

From these last two facts, and the M-Weierstrass test for uniform convergence of series of functions, we have that both $\sum_{n=1}^{\infty} a_n \cos nx$ and $\sum_{n=1}^{\infty} b_n \sin nx$ converge uniformly on $[-\pi, \pi]$, since for all n

$$|a_n \cos nx| \leq |a_n|, \qquad |b_n \sin nx| \leq |b_n|.$$

Thus the Fourier series of f converges uniformly on $[-\pi, \pi]$. Since $f' \in E$ we have that f satisfies the conditions of Dirichlet's Theorem and thus the Fourier series of f converges pointwise to f at every point in $[-\pi, \pi]$. This being so, it converges uniformly to f on the full interval. ∎

Remark: In the course of the proof of Theorem 2.15 we came across the seeds of a phenomenon important in the theory of Fourier series. There is a correlation of sorts between the smoothness of a function f (as a 2π-periodic function) and the rate at which its Fourier coefficients converge to zero. It follows from the proof of Theorem 2.15 that if $f, f', \ldots, f^{(k-1)}$ are continuous 2π-periodic functions and $f^{(k)} \in E$, then $n^k a_n$ and $n^k b_n$ both tend to zero as n tends to ∞. There are converse results of a similar nature. Why should smoothness be related to the rate at which the coefficients tend to zero? After all, each $\sin nx$ and $\cos nx$ is infinitely smooth. Heuristically, when dealing with Fourier series, non-smoothness comes in using those smooth functions which oscillate more and more, i.e., $\sin nx$ and $\cos nx$ for large n.

We now consider the somewhat more delicate question of the uniform convergence of Fourier series on subintervals $[a, b]$ of $[-\pi, \pi]$, in the case where f has points of discontinuity. Obviously, if the subinterval $[a, b]$ contains jump points of f then the Fourier series of f does not converge uniformly to f on $[a, b]$. We will prove that if $f' \in E$ and if the subinterval $[a, b]$ does not contain any jump points of f, then the Fourier series of f converges uniformly to f on $[a, b]$. We start with a specific example. We will then use this example to prove the general result. Let $\phi(x) = x$, for $-\pi < x < \pi$, and $\phi(-\pi) = \phi(\pi) = 0$. We further assume that ϕ is defined on all of $\mathbb{R}$ and is 2π-periodic. From Example 2.1 and Dirichlet's Theorem (Theorem 2.6) it follows that for all real x we have the equality

$$\phi(x) = \sum_{n=1}^{\infty} \frac{2(-1)^{n+1}}{n} \sin nx.$$

That is, the Fourier series converges pointwise to ϕ on all of $[-\pi, \pi]$. However, this convergence is not uniform as ϕ is not continuous at $x = \pm\pi$. In this next proposition we prove that the Fourier series of ϕ uniformly converges to ϕ on every closed subinterval of $(-\pi, \pi)$ (which does not include the points $\pm\pi$).

Proposition 2.16: *For every $0 < b < \pi$, the Fourier series of ϕ converges uniformly to ϕ on $[-b, b]$.*

Proof: We will use the Cauchy criteria for uniform convergence. As usual, set

$$S_m(x) = \sum_{n=1}^{m} \frac{2(-1)^{n+1}}{n} \sin nx.$$

Let $\epsilon > 0$. We must prove that there exists an $N(\epsilon)$ such that $|S_m(x) - S_k(x)| < \epsilon$ for all $m, k \geq N(\epsilon)$ and every $x \in [-b, b]$. To this end we first consider the function $\cos \frac{x}{2}$ on the interval $[-b, b]$. This function attains its minimum value $\cos \frac{b}{2} > 0$ at the endpoints $\pm b$. Let m and k be two natural numbers and assume that $m > k$. Then

$$|S_m(x) - S_k(x)| = 2 \left| \frac{\sin(k+1)x}{k+1} - \frac{\sin(k+2)x}{k+2} + \cdots \pm \frac{\sin mx}{m} \right|.$$

We multiply the right-hand side of the above equation by $\cos \frac{x}{2}$ and use the trigonometric identity

$$2 \sin \alpha \cos \beta = \sin(\alpha + \beta) + \sin(\alpha - \beta).$$

Thus

$$2\cos\frac{x}{2} \left[\frac{\sin(k+1)x}{k+1} - \frac{\sin(k+2)x}{k+2} + \cdots \pm \frac{\sin mx}{m} \right] =$$

$$= \left[\frac{\sin(k+\frac{3}{2})x}{k+1} + \frac{\sin(k+\frac{1}{2})x}{k+1} \right] - \left[\frac{\sin(k+\frac{5}{2})x}{k+2} + \frac{\sin(k+\frac{3}{2})x}{k+2} \right]$$

$$+ \cdots \pm \left[\frac{\sin(m+\frac{1}{2})x}{m} + \frac{\sin(m-\frac{1}{2})x}{m} \right].$$

Let $c = \cos \frac{b}{2}$. Since $\cos \frac{x}{2} \geq c$ for all $x \in [-b, b]$, and because $|\sin \alpha| \leq 1$ for all real α, we obtain after a little calculation that for all $x \in [-b, b]$

$$|S_m(x) - S_k(x)| \leq \frac{1}{c} \left[\frac{1}{k+1} + \frac{1}{(k+1)(k+2)} + \frac{1}{(k+2)(k+3)} + \cdots + \frac{1}{(m-1)m} + \frac{1}{m} \right].$$

The series $\sum_{n=1}^{\infty} \frac{1}{n(n+1)}$ converges. Thus there exists a natural number $N(\epsilon)$ such that for all $k \geq N(\epsilon)$,

$$\sum_{n=k}^{\infty} \frac{1}{n(n+1)} < \frac{c\epsilon}{2}.$$

In addition we may assume that for all $r \geq N(\epsilon)$ we have $\frac{1}{r} < \frac{c\epsilon}{4}$. Thus for all $m > k \geq N(\epsilon)$ and for all $x \in [-b, b]$,

$$|S_m(x) - S_k(x)| \leq \frac{1}{c}\left[\frac{1}{k+1} + \sum_{n=k}^{m} \frac{1}{n(n+1)} + \frac{1}{m}\right] < \frac{1}{c}\left[\frac{c\epsilon}{4} + \frac{c\epsilon}{2} + \frac{c\epsilon}{4}\right] = \epsilon.$$

The sequence of functions $\{S_m\}_{m=1}^{\infty}$ satisfies the Cauchy criterion for uniform convergence. It therefore converges uniformly to the function $\phi(x) = x$ on the interval $[-b, b]$. ∎

We can now prove our main result on this subject.

Theorem 2.17: *Let the function f be such that $f, f' \in E$. Assume that*

$$-\pi < d_1 < d_2 < \cdots < d_n \leq \pi$$

are the jump points of f in $[-\pi, \pi]$. If $[a, b]$ is a subinterval of $[-\pi, \pi]$ which does not contain any of the points d_k, then the Fourier series of f converges uniformly to f on $[a, b]$.

Proof: For every $k \in \{1, \ldots, n\}$, set $z_k = f(d_k+) - f(d_k-)$. We define the function

$$g(x) = f(x) + \frac{z_1}{2\pi}\phi(x + \pi - d_1) + \frac{z_2}{2\pi}\phi(x + \pi - d_2) + \cdots + \frac{z_n}{2\pi}\phi(x + \pi - d_n)$$

where ϕ is as defined before Proposition 2.16. It is easily verified that g is continuous on $[-\pi, \pi]$, except possibly at the points d_k. However, the discontinuity at these points is removable (since the one-sided limits at each point exist and are equal). Thus, we may assume that g is continuous on $[-\pi, \pi]$ and $g(-\pi) = g(\pi)$. In addition $g' \in E$. Therefore by Theorem 2.15 the Fourier series of g converges uniformly to g on $[-\pi, \pi]$. From Proposition 2.16 it follows that the Fourier series of ϕ converges uniformly to ϕ on any closed subinterval of $[-\pi, \pi]$ which does not include the points $x = \pm\pi$. Thus the Fourier series of $\phi(x + \pi - d_k)$ (which we obtain by a simple shift of the Fourier series of ϕ) converges uniformly on any closed subinterval of $[-\pi, \pi]$ which does not contain the point $x = d_k$. This fact is of course also true for the function $\frac{z_k}{2\pi}\phi(x + \pi - d_k)$. The interval $[a, b]$ does not contain any of the points $x = d_k$, $1 \leq k \leq n$. This being so, each of the Fourier series of the functions $\frac{z_k}{2\pi}\phi(x + \pi - d_k)$ converges uniformly on $[a, b]$. Since

$$f(x) = g(x) - \left[\frac{z_1}{2\pi}\phi(x + \pi - d_1) + \cdots + \frac{z_n}{2\pi}\phi(x + \pi - d_n)\right],$$

the Fourier series of f (which we obtain by subtracting the Fourier series of each of the $\frac{z_k}{2\pi}\phi(x + \pi - d_k)$ from the Fourier series of g) converges uniformly to f on $[a, b]$. ∎

Exercises

1. Let
$$f(x) = \begin{cases} Ax + B, & -\pi \le x < 0, \\ \cos x, & 0 \le x \le \pi. \end{cases}$$
   For what values A and B does the Fourier series of f converge uniformly to f on all of $[-\pi, \pi]$?

2. Let $g(x) = \begin{cases} \cos x, & -\pi < x < 0, \\ \sin x, & 0 < x < \pi. \end{cases}$

   (a) Calculate the Fourier series of g.

   (b) Define the function
   $$h(x) = \int_{-\pi}^{x} g(t)\, dt + a \sin \frac{x}{2}, \qquad -\pi \le x \le \pi.$$

   For what values of a does the Fourier series of h converge uniformly to h on $[-\pi, \pi]$?

3. Prove that for all $-\pi \le x \le \pi$ and all $a \notin \mathbb{Z}$
$$\cos ax = \frac{\sin \pi a}{\pi a} + \sum_{n=1}^{\infty} (-1)^n \frac{2a \sin \pi a}{\pi(a^2 - n^2)} \cos nx.$$
   Using this fact show that
$$\cot \pi a = \frac{1}{\pi}\left[\frac{1}{a} - \sum_{n=1}^{\infty} \frac{2a}{n^2 - a^2}\right].$$

4. Let $f \in E$ and assume
$$\frac{a_0}{2} + \sum_{n=1}^{\infty} [a_n \cos nx + b_n \sin nx]$$
   is the Fourier series of f. Show that if there exist constants c and d such that
$$|a_n| \le \frac{c}{n^2}, \qquad |b_n| \le \frac{d}{n^2}$$
   for all n, then f may be considered to be continuous on $[-\pi, \pi]$, satisfying $f(-\pi) = f(\pi)$, and the Fourier series of f converges uniformly to f on $[-\pi, \pi]$.

5. Assume that f is continuous on $[-\pi, \pi]$ and $f(-\pi) = f(\pi)$. Prove that f can be uniformly approximated by trigonometric polynomials. That is,

given $\epsilon > 0$ there exists a trigonometric polynomial T of some degree such that
$$|f(x) - T(x)| < \epsilon$$
for all $x \in [-\pi, \pi]$.

6. Parseval's Identity

In this section we prove that the linear space E is closed with respect to the Fourier series and associated norm. This fact is equivalent (see Proposition 1.24) to Parseval's identity for all $f \in E$. We state the theorem in terms of Parseval's identity.

Theorem 2.18: (*Parseval's Identity*) *For each $f \in E$ we have the equality*

(2.8)
$$\frac{1}{\pi} \int_{-\pi}^{\pi} |f(x)|^2 dx = \frac{|a_0|^2}{2} + \sum_{n=1}^{\infty} \left(|a_n|^2 + |b_n|^2\right)$$

where the a_n and b_n are the Fourier coefficients of f.

In the discussion of Proposition 2.10 (Bessel's inequality) we saw why (2.8) is the correct form of Parseval's identity, as given in Proposition 1.24, for the special case of Fourier series. As was noted in Proposition 1.24, Parseval's identity is equivalent to the property of closure. To prove Parseval's identity it suffices to prove that for each $f \in E$ we have

(2.9)
$$\lim_{m \to \infty} \frac{1}{\pi} \int_{-\pi}^{\pi} |f(x) - S_m(x)|^2 dx = 0.$$

We will prove this result in two stages. At the first stage we prove that (2.9) holds for all functions f satisfying the conditions of Theorem 2.15. At the second stage we show how it is possible to approximate each $f \in E$ by functions satisfying the conditions of Theorem 2.15. The result then follows.

Proposition 2.19: *If f is continuous in $[-\pi, \pi]$, $f(-\pi) = f(\pi)$, and $f' \in E$, then*
$$\lim_{m \to \infty} \|f - S_m\| = 0.$$

Proof: From Theorem 2.15, the Fourier series of f converges uniformly to f on $[-\pi, \pi]$. Thus for each given $\epsilon > 0$ there exists $N(\epsilon)$ such that for each $x \in [-\pi, \pi]$ and for all $m \geq N(\epsilon)$
$$|f(x) - S_m(x)| < \epsilon.$$

Therefore for every $m \geq N(\epsilon)$

$$\|f - S_m\|^2 = \frac{1}{\pi}\int_{-\pi}^{\pi} |f(x) - S_m(x)|^2 dx \leq 2\epsilon^2.$$

That is to say, $\lim_{m\to\infty} \|f - S_m\|^2 = 0$. ■

Proposition 2.20: *Let $f \in E$ and $\epsilon > 0$. There exists a function g which is continuous on $[-\pi, \pi]$, $g(-\pi) = g(\pi)$, $g' \in E$, and $\|f - g\| < \epsilon$.*

Proof: If we consider the definition of the Riemann integral we will find that for each $f \in E$ and $\epsilon > 0$ there exist $n+1$ points (n depends on ϵ)

$$-\pi = d_0 < d_1 < \cdots < d_n = \pi$$

and n values c_k, $1 \leq k \leq n$, such that the step function

$$h(x) = c_k, \qquad d_{k-1} < x \leq d_k, \quad k = 1, 2, \ldots, n,$$

satisfies

$$\|f - h\| < \frac{\epsilon}{2}.$$

A visualization of this is given in Figure 2.2.

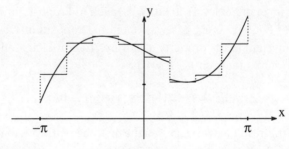

Fig. 2.2

In addition, if $|f(x)| \leq M$ for all x, then we can choose the values c_k, $1 \leq k \leq n$, such that $|h(x)| \leq M$ for all x. The idea now is to alter h so that it satisfies the conditions of the proposition. To this end we choose $\delta > 0$ which satisfies $2\delta < d_k - d_{k-1}$, $1 \leq k \leq n$, and $0 < \delta < \frac{\epsilon^2 \pi}{32nM^2}$. We now define the function g by

$$g(x) = \begin{cases} c_k, & d_{k-1} + \delta \leq x \leq d_k - \delta, \quad k = 1, 2, \ldots, n, \\ \frac{c_{k+1} - c_k}{2\delta}(x - d_k) + \frac{c_{k+1} + c_k}{2}, & d_k - \delta < x < d_k + \delta, \; k = 1, \ldots, n-1, \\ \frac{c_1 - c_n}{2\delta}(x - \pi) + \frac{c_1 + c_n}{2}, & \pi - \delta < x < \pi + \delta, \end{cases}$$

Chapter 2: Fourier Series

and such that g is 2π-periodic. It is easily checked that g is continuous on $[-\pi, \pi]$, $g(-\pi) = g(\pi)$, and $g' \in E$. In addition

$$\|h - g\|^2 = \frac{1}{\pi} \int_{-\pi}^{\pi} |h(x) - g(x)|^2 dx$$

$$= \sum_{k=1}^{n-1} \frac{1}{\pi} \int_{d_k-\delta}^{d_k+\delta} |h(x) - g(x)|^2 dx + \frac{1}{\pi} \int_{-\pi}^{-\pi+\delta} |h(x) - g(x)|^2 dx$$

$$+ \frac{1}{\pi} \int_{\pi-\delta}^{\pi} |h(x) - g(x)|^2 dx$$

$$\leq \frac{4M^2}{\pi} \left[\sum_{k=1}^{n-1} \int_{d_k-\delta}^{d_k+\delta} dx + \int_{-\pi}^{-\pi+\delta} dx + \int_{\pi-\delta}^{\pi} dx \right]$$

$$= \frac{8nM^2\delta}{\pi}$$

$$< \left(\frac{\epsilon}{2}\right)^2$$

and therefore
$$\|f - g\| \leq \|f - h\| + \|h - g\| < \frac{\epsilon}{2} + \frac{\epsilon}{2} = \epsilon. \qquad \blacksquare$$

Proof of Theorem 2.18: Let $f \in E$ and $\epsilon > 0$. By Proposition 2.20 there exists a function g satisfying the conditions of Proposition 2.20 such that

$$\|f - g\| < \epsilon.$$

In addition, there exists $N(\epsilon)$ such that for all $m \geq N(\epsilon)$,

$$\|g - T_m\| < \epsilon$$

where T_m is the mth partial sum of the Fourier series of g. Thus for $m \geq N(\epsilon)$ we have

$$\|f - T_m\| \leq \|f - g\| + \|g - T_m\| < 2\epsilon.$$

By Theorem 2.3
$$\|f - S_m\| \leq \|f - T_m\|$$

where S_m is the mth partial sum of the Fourier series of f. Thus for all $m \geq N(\epsilon)$, $\|f - S_m\| < 2\epsilon$, and hence $\lim_{m \to \infty} \|f - S_m\| = 0$. $\qquad \blacksquare$

Example 2.9: In Example 2.5 we calculated the Fourier series of $f(x) = x^2$:

$$x^2 \sim \frac{\pi^2}{3} + \sum_{n=1}^{\infty} \frac{4(-1)^n}{n^2} \cos nx.$$

From Parseval's identity we obtain
$$\frac{1}{\pi}\int_{-\pi}^{\pi} x^4 dx = \frac{2\pi^4}{9} + \sum_{n=1}^{\infty} \frac{16}{n^4}.$$

Since $\frac{1}{\pi}\int_{-\pi}^{\pi} x^4 dx = \frac{2\pi^4}{5}$, it follows that
$$\sum_{n=1}^{\infty} \frac{1}{n^4} = \frac{\pi^4}{90}.$$

We have obtained the calculation of a numerical series by rather elementary means. Similar series can also be calculated from this series. For example
$$\sum_{k=1}^{\infty} \frac{1}{(2k-1)^4} = \sum_{n=1}^{\infty} \frac{1}{n^4} - \sum_{k=1}^{\infty} \frac{1}{(2k)^4} = \sum_{n=1}^{\infty} \frac{1}{n^4} - \frac{1}{16}\sum_{k=1}^{\infty} \frac{1}{k^4}$$
$$= \frac{15}{16}\sum_{n=1}^{\infty} \frac{1}{n^4} = \frac{15}{16} \cdot \frac{\pi^4}{90} = \frac{\pi^4}{96}.$$

An immediate consequence of the two propositions and theorem of this section is that the sequence $\{e^{inx}\}_{n=-\infty}^{\infty}$ is also a closed orthonormal system in E. The associated Parseval identity in this case is
$$\frac{1}{2\pi}\int_{-\pi}^{\pi} |f(x)|^2 dx = \sum_{n=-\infty}^{\infty} |c_n|^2$$
where $f \in E$ and $\sum_{n=-\infty}^{\infty} c_n e^{inx}$ is the complex Fourier series of f.

Example 2.10: Consider the complex Fourier series of $f(x) = e^x$. For each $n \in \mathbb{Z}$
$$c_n = \frac{1}{2\pi}\int_{-\pi}^{\pi} e^x e^{-inx} dx = \frac{1}{2\pi}\int_{-\pi}^{\pi} e^{(1-in)x} dx = \frac{1}{2\pi} \frac{e^{(1-in)x}}{1-in}\bigg|_{-\pi}^{\pi}$$
$$= \frac{(-1)^n (e^\pi - e^{-\pi})}{2\pi(1-in)}$$

and thus
$$e^x \sim \sum_{n=-\infty}^{\infty} \frac{(-1)^n (e^\pi - e^{-\pi})}{2\pi(1-in)} e^{inx}.$$

A simple calculation gives $\frac{1}{2\pi}\int_{-\pi}^{\pi} |f(x)|^2 dx = \frac{e^{2\pi} - e^{-2\pi}}{4\pi}$. Thus by the complex Parseval identity we obtain
$$\frac{e^{2\pi} - e^{-2\pi}}{4\pi} = \sum_{n=-\infty}^{\infty} \frac{(e^\pi - e^{-\pi})^2}{4\pi^2 |1-in|^2} = \frac{(e^\pi - e^{-\pi})^2}{4\pi^2} \sum_{n=-\infty}^{\infty} \frac{1}{n^2+1}$$

and therefore
$$\sum_{n=-\infty}^{\infty} \frac{1}{n^2+1} = \frac{\pi(e^\pi + e^{-\pi})}{e^\pi - e^{-\pi}} = \pi \coth \pi.$$

We now present another property of Fourier series which is a generalization of Parseval's identity.

Proposition 2.21: (**Generalized Parseval Identity**) *For each $f, g \in E$,*
$$\frac{1}{\pi}\int_{-\pi}^{\pi} f(x)\overline{g(x)}\,dx = \frac{a_0 \overline{c_0}}{2} + \sum_{n=1}^{\infty}\left(a_n \overline{c_n} + b_n \overline{d_n}\right)$$
where
$$f(x) \sim \frac{a_0}{2} + \sum_{n=1}^{\infty}[a_n \cos nx + b_n \sin nx]$$
and
$$g(x) \sim \frac{c_0}{2} + \sum_{n=1}^{\infty}[c_n \cos nx + d_n \sin nx].$$

The claim is an immediate consequence of Proposition 1.27. If we set $f = g$ in this proposition we obtain Parseval's identity.

We end this section by presenting a consequence of Parseval's identity. We show that if two functions in E have the same Fourier series then they are essentially identical. If these two functions are in E' then this claim is an immediate consequence of Dirichlet's Theorem (Theorem 2.6). With respect to functions which only belong to E we need Parseval's identity to prove this claim.

Proposition 2.22: (**Uniqueness of Fourier Series**) *If $f, g \in E$ and the Fourier series of f and g are identical, then $f(x) = g(x)$ except at a finite number of points.*

Proof: The coefficients in the Fourier series of $f - g \in E$ are all identically zero. From Parseval's identity
$$\|f - g\| = 0.$$

From the properties of the norm it follows that $f = g$ except at a finite number of points (see Remark (e) after Definition 2.2). ∎

Exercises

1. Calculate the integral
$$\int_{-\pi}^{\pi} \left| \sum_{n=1}^{\infty} \frac{1}{2^n} e^{inx} \right|^2 dx.$$

2. Let $f \in E$ and
$$\frac{a_0}{2} + \sum_{n=1}^{\infty}[a_n \cos nx + b_n \sin nx]$$
denote the Fourier series of f on $[-\pi, \pi]$. Determine the value of
$$\frac{1}{\pi}\int_{-\pi}^{\pi} |f(x+\pi) - f(x)|^2 dx$$
in terms of the a_n and b_n.

3. For each natural integer n we define
$$f_n(x) = 1 + \sum_{k=1}^{n}[\cos kx - \sin kx].$$
Calculate the value of the integral $\int_{-\pi}^{\pi} |f_n(x)|^2 dx$.

4. Let $f \in E$ and
$$f(x) \sim \frac{a_0}{2} + \sum_{n=1}^{\infty}[a_n \cos nx + b_n \sin nx]$$
denote the Fourier series of f. Calculate $I = \frac{a_0^2}{2} + \sum_{n=1}^{\infty} a_n^2$, when $f(x) = e^{-x}$.

7. The Gibbs Phenomenon

In Section 2.5 we were able to formulate sufficient conditions under which the Fourier series converges uniformly on $[-\pi, \pi]$ or subintervals thereof to its function. Uniform convergence is the best possible convergence. However, it does not always hold. In this section we consider situations in which we do not have uniform convergence.

Let f be a 2π-periodic function satisfying the conditions of Dirichlet's Theorem on $[-\pi, \pi]$. Let
$$-\pi < d_1 < d_2 < \cdots < d_n \leq \pi$$
denote the jump points of f in $[-\pi, \pi]$. In a previous section (see Theorem 2.17) we proved that under certain conditions the Fourier series of f converges uniformly on every subinterval $[a, b]$ of $[-\pi, \pi]$ which does not contain any of these points. But at the points d_k, $1 \leq k \leq n$, something rather odd occurs which is called the "Gibbs phenomenon".

This phenomenon was noted by the physicist A. Michelson at the end of the nineteenth century. He built a "machine" (a harmonic analyser and synthesizer) which could calculate some initial Fourier coefficients of a graphically given function f, and then also graphically "exhibit" the function from these coefficients (i.e., graphically represent some partial sums S_m). Michelson found that for "good" functions (those satisfying the conditions of Theorem 2.15) the graphs of the S_m were quite close to the function f. However, for the function $f(x) = \text{sgn}(x)$ the graph of the partial sums exhibited a large error in the neighbourhood of $x = 0$ and $x = \pm\pi$ (the jumps of the function) independent of the number of terms in the partial sum. The explanation for this phenomenon was provided by J. W. Gibbs who showed that the errors were not the result of machine error, and explained their source. This being so, this phenomenon was given his name. (In fact the phenomenon had been discovered some 60 years previously by a mathematician named Wilbraham but forgotten.) What is interesting is that the same relative error occurs at every jump discontinuity for every $f \in E'$. To explain this phenomenon we start with an example.

Example 2.11: Let $\phi(x) = x$, $-\pi < x < \pi$, and $\phi(-\pi) = \phi(\pi) = 0$. As usual we assume that ϕ is 2π-periodic and defined on all of $\mathbb{R}$. In Example 2.1 we calculated the Fourier series of ϕ and found it to be

$$\sum_{n=1}^{\infty} \frac{2(-1)^{n+1}}{n} \sin nx.$$

In Proposition 2.16 we proved that this Fourier series converges uniformly to ϕ on $[-b, b]$ for every $0 < b < \pi$. Since ϕ is not continuous as a 2π-periodic function, we know that its Fourier series does not converge uniformly on all of $[-\pi, \pi]$. What is happening in the neighbourhoods of $x = \pm\pi$?

For each natural number m, let

$$T_m(x) = \sum_{n=1}^{m} \frac{2(-1)^{n+1}}{n} \sin nx$$

denote the partial sum of the Fourier series of ϕ. We will set $x_m = \pi - \frac{\pi}{m}$ and approximately calculate $T_m(x_m)$. We have

$$T_m(x_m) = \sum_{n=1}^{m} \frac{2(-1)^{n+1}}{n} \sin n(\pi - \tfrac{\pi}{m}) = \sum_{n=1}^{m} \frac{2}{n} \sin \tfrac{n\pi}{m}.$$

We rewrite this as

$$T_m(x_m) = \sum_{n=1}^{m} \frac{2}{n} \sin \tfrac{n\pi}{m} = 2 \sum_{n=1}^{m} \frac{\sin \tfrac{n\pi}{m}}{\tfrac{n\pi}{m}} \frac{\pi}{m}.$$

The last expression is supposed to convince you of the fact that $T_m(x_m)$ is essentially a Riemann sum of the integral

$$2 \int_0^{\pi} \frac{\sin x}{x} \, dx$$

obtained by dividing $[0, \pi]$ into the m subintervals $\left[\frac{(n-1)\pi}{m}, \frac{n\pi}{m}\right]$, $1 \leq n \leq m$. Each such subinterval is of length $\frac{\pi}{m}$, and in each such subinterval we chose to evaluate the function at the right-hand endpoint $\frac{n\pi}{m}$, $1 \leq n \leq m$. Thus

$$\lim_{m \to \infty} T_m(x) = 2 \int_0^\pi \frac{\sin x}{x}\, dx \approx 1.18\pi.$$

The point x_m, as m tends to infinity, approaches $x = \pi$ from the left. Thus $\phi(x_m)$ approaches the value π. The jump at the point $x = \pi$ is $\phi(\pi-) - \phi(\pi+) = 2\pi$, and thus for m sufficiently large

$$\frac{T_m(x_m) - \phi(x_m)}{\phi(\pi-) - \phi(\pi+)} \approx \frac{1.18\pi - \pi}{2\pi} = 0.09.$$

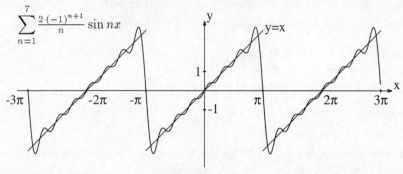

Fig. 2.3

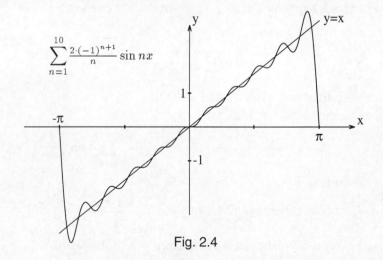

Fig. 2.4

In Figures 2.3–2.4 we have overlaid the graphs of T_7 and T_{10}, respectively, on that of ϕ. Note what happens in a neighbourhood of $x = \pm\pi$ (and $x = \pm 3\pi$).

Chapter 2: Fourier Series

The graphs of T_7 and T_{10} are at some distance from ϕ in the opposite direction to the jump. (That is called "overshooting".)

On the basis of this last example, we present a general proof for the existence of the Gibbs phenomenon for functions satisfying the conditions of Theorem 2.17. The idea behind the proof of Proposition 2.23 is essentially the same as that in the proof of Theorem 2.17. We subtract away from our "problem" function f a suitable shift and multiple of the above ϕ. This new function and its Fourier series are now very well behaved. This being so, the bad behaviour of f and the bad behaviour of ϕ have cancelled each other out, and thus are equivalent.

Proposition 2.23: *Let f be a function for which $f, f' \in E$. Let d be a point of discontinuity of f, $-\pi \leq d < \pi$, and*

$$S_m(x) = \frac{a_0}{2} + \sum_{n=1}^{m}[a_n \cos nx + b_n \sin nx]$$

be the partial sum of the Fourier series of f. Then there exists a sequence of points $\{x_m\}_{m=1}^{\infty}$, $x_m > d$, such that $\lim_{m \to \infty} x_m = d$ and

$$\lim_{m \to \infty} \frac{S_m(x_m) - f(x_m)}{f(d+) - f(d-)} \geq 0.089.$$

Proof: Assume for convenience that $-\pi < d < \pi$, and let

$$z = f(d+) - f(d-).$$

Let ϕ be the function of Example 2.11 and T_m the partial sums of the Fourier series of ϕ, as defined therein. Let us define

$$g(x) = f(x) + \frac{z}{2\pi}\phi(x + \pi - d).$$

It follows that $g(d-) = g(d+)$ (see the same procedure in the proof of Theorem 2.17). Now $g, g' \in E$ and thus we can assume that g is continuous at d. This being so, there exists a sufficiently small $\delta > 0$ such that g is continuous on the interval $[d - \delta, d + \delta]$. From Theorem 2.17, the Fourier series of g converges uniformly to g in $[d - \delta, d + \delta]$. Let U_m denote the partial sums of the Fourier series of g. From the definition of g we have

$$f(x) = g(x) - \frac{z}{2\pi}\phi(x + \pi - d)$$

from which it follows that for all $x \in [-\pi, \pi]$

$$S_m(x) = U_m(x) - \frac{z}{2\pi}T_m(x + \pi - d).$$

Let $\epsilon > 0$. Since U_m converges uniformly to g on $[d - \delta, d + \delta]$, there exists an m_1 such that for all $m \geq m_1$, and all $x \in [d - \delta, d + \delta]$, we have

(2.10) $$|g(x) - U_m(x)| < \epsilon |z|.$$

From Example 2.11 there exists an $m_0 \geq m_1$, such that for all $m \geq m_0$ we have a point x_m in the interval $[d, d + \delta]$ satisfying

(2.11) $$\frac{\phi(x_m + \pi - d) - T_m(x_m + \pi - d)}{2\pi} \geq 0.089.$$

(0.089 is slightly smaller than the real value which can be calculated in Example 2.11). Thus for all $m \geq m_0$ it follows from (2.10) and (2.11) that

$$\frac{S_m(x_m) - f(x_m)}{f(d+) - f(d-)} \geq 0.089,$$

which proves the claim. ∎

Remark: In the same way we can prove the existence of $\tilde{x}_m$ in the interval $[d - \delta, d]$ for which
$$\frac{S_m(\tilde{x}_m) - f(\tilde{x}_m)}{f(d+) - f(d-)} \leq -0.089.$$

We did not prove that the difference between S_m and f in a neighbourhood of d will not be larger. However, this is essentially true. The phenomenon which occurs at each jump discontinuity of f (satisfying $f, f' \in E$) is exactly that which occurs for the function $\phi(x) = x$ in a neighbourhood of the points $x = \pm\pi$.

8. Sine and Cosine Series

Let f be a function satisfying the conditions of Dirichlet's Theorem on the interval $[0, \pi]$. There exists an infinite number of totally different "Fourier series" which converge to f pointwise on $[0, \pi]$. Does this seem strange? Well the reason for this is that we may extend the function f to the interval $[-\pi, \pi]$ in an infinite number of different ways (i.e., define f on $[-\pi, 0)$) such that the extension satisfies the conditions of Dirichlet's Theorem on $[-\pi, \pi]$. For each such extension there exists an appropriate Fourier series which, on the interval $(0, \pi)$, converges to f pointwise (in the manner given by Dirichlet's Theorem). Among all possible extensions of f to $[-\pi, \pi]$ there are a few worthy of mention which lead to simple and useful series. We consider two of them in this section.

In Section 2.2 we noted that if g is an even function on $[-\pi, \pi]$, then the Fourier series of g contains only cosine terms, i.e., all $b_n = 0$. Similarly, if g

Chapter 2: Fourier Series

is odd then the Fourier series of g contains only sine terms, i.e., all $a_n = 0$. Such series are, in a sense, simpler than the general series. This fact leads us to define two different series which represent f on the interval $[0, \pi]$.

Let f be a given function defined on the interval $[0, \pi]$. We define the function $\hat{f}$ on $[-\pi, \pi]$ by

$$\hat{f}(x) = \begin{cases} f(-x), & -\pi \leq x < 0, \\ f(x), & 0 \leq x < \pi. \end{cases}$$

Thus $\hat{f}$ is an even function on $[-\pi, \pi]$ and is called the *even extension* of f. From Proposition 2.4, the Fourier series of $\hat{f}$ is

$$\hat{f}(x) \sim \frac{a_0}{2} + \sum_{n=1}^{\infty} a_n \cos nx$$

where

$$a_n = \frac{2}{\pi} \int_0^{\pi} f(x) \cos nx \, dx.$$

Similarly we define the function $\tilde{f}$ on $[-\pi, \pi]$ by

$$\tilde{f}(x) = \begin{cases} -f(-x), & -\pi \leq x < 0, \\ 0, & x = 0, \\ f(x), & 0 < x \leq \pi. \end{cases}$$

$\tilde{f}$ is an odd function on $[-\pi, \pi]$ and is called the *odd extension* of f. Again, by Proposition 2.4, the Fourier series of $\tilde{f}$ is

$$\tilde{f}(x) \sim \sum_{n=1}^{\infty} b_n \sin nx$$

where

$$b_n = \frac{2}{\pi} \int_0^{\pi} f(x) \sin nx \, dx.$$

Definition 2.24: *Let f be a piecewise continuous function defined on $[0, \pi]$. The series*

$$\frac{a_0}{2} + \sum_{n=1}^{\infty} a_n \cos nx$$

where

$$a_n = \frac{2}{\pi} \int_0^{\pi} f(x) \cos nx \, dx$$

is called the cosine series of f. Similarly, the series

$$\sum_{n=1}^{\infty} b_n \sin nx$$

where

$$b_n = \frac{2}{\pi} \int_0^{\pi} f(x) \sin nx \, dx$$

is called the sine series of f.

Let us consider some theoretical consequences. Let $E[0, \pi]$ denote the complex-valued piecewise continuous functions defined on $[0, \pi]$. On $E[0, \pi]$ we define the inner product

$$\langle f, g \rangle = \frac{2}{\pi} \int_0^{\pi} f(x) \overline{g(x)} dx.$$

It is not difficult to check that each of the series $\{\sin nt\}_{n=1}^{\infty}$ and $\{\cos nt\}_{n=1}^{\infty} \cup \{\frac{1}{\sqrt{2}}\}$ represents an orthonormal system on the space $E[0, \pi]$. From the above discussion it actually follows that each of the above orthonormal systems is closed. This property is equivalent to Parseval's identity holding for all $f \in E[0, \pi]$ (see Proposition 1.24). For the cosine series Parseval's identity becomes

$$\frac{2}{\pi} \int_0^{\pi} |f(x)|^2 dx = \frac{|a_0|^2}{2} + \sum_{n=1}^{\infty} |a_n|^2$$

where $a_n = \frac{2}{\pi} \int_0^{\pi} f(x) \cos nx \, dx$. For the sine series Parseval's identity is

$$\frac{2}{\pi} \int_0^{\pi} |f(x)|^2 dx = \sum_{n=1}^{\infty} |b_n|^2$$

where $b_n = \frac{2}{\pi} \int_0^{\pi} f(x) \sin nx \, dx$. Another result is the following consequence of Theorem 2.15.

Proposition 2.25: (a) *If f is continuous on $[0, \pi]$, $f(0) = f(\pi) = 0$, and $f' \in E[0, \pi]$, then the sine series of f converges uniformly to f on $[0, \pi]$.*

(b) *If f is continuous on $[0, \pi]$ and $f' \in E[0, \pi]$, then the cosine series of f converges uniformly to f on $[0, \pi]$.*

In (a) the condition $f(0) = f(\pi) = 0$ is necessary for the uniform convergence (every sine series necessarily vanishes at $x = 0$ and $x = \pi$). In (b), on the other hand, there is no restriction on the endpoint values.

Exercises

1. Let
$$f(x) = \begin{cases} x, & 0 \le x \le \frac{\pi}{2}, \\ \pi - x, & \frac{\pi}{2} \le x \le \pi, \end{cases}$$

   and let $\tilde{f}$ be the odd extension of f to $[-\pi, \pi]$. Find the Fourier series of $\tilde{f}$ on $[-\pi, \pi]$.

2. Let $f \in E[0, \pi]$ and
$$f(x) \sim \sum_{n=1}^{\infty} b_n \sin nx$$

   denote the sine series of f, while
$$f(x) \sim \frac{a_0}{2} + \sum_{n=1}^{\infty} a_n \cos nx$$

   denotes the cosine series of f. What equals the function
$$g(x) = \frac{a_0}{2} + \sum_{n=1}^{\infty} [a_n \cos nx + b_n \sin nx]$$

   at each point in $[-\pi, \pi]$?

3. (a) Find the sine series of the function $f(x) = x(\pi - x)$ defined on $[0, \pi]$.
   (b) Prove that
$$\sum_{n=1}^{\infty} \frac{1}{n^6} = \frac{\pi^6}{945}.$$
   (c) Prove that
$$\frac{\pi^3}{32} = \frac{1}{1^3} - \frac{1}{3^3} + \frac{1}{5^3} - \frac{1}{7^3} + \cdots.$$

4. (a) Find the sine series of the function $\cos x$ on the interval $[0, \pi]$.
   (b) To what does the series converge for each point in $[-\pi, \pi]$?
   (c) Define the function
$$g(x) = \sum_{n=1}^{\infty} b_n \sin nx.$$

   Sketch the graph of g on $[-\pi, \pi]$.

9. Differentiation and Integration of Fourier Series

One of the fundamental problems of mathematical analysis has to do with the infinite in its various modes. We can find examples of this in almost every section of this text. One problem which exists in this connection has to do with the interchange of orders of operation (like integration, differentiation, infinite summing, etc...). In this section we consider two such problems.

Let f be a function for which $f, f' \in E$, with Fourier series

$$f(x) \sim \frac{a_0}{2} + \sum_{n=1}^{\infty} [a_n \cos nx + b_n \sin nx].$$

Is it true that

$$f'(x) \sim \sum_{n=1}^{\infty} [-na_n \sin nx + nb_n \cos nx] \,?$$

In other words, can we differentiate term by term the Fourier series of a function f in order to obtain the Fourier series of f'? In general the answer to this question is no. Let us consider a simple example which illustrates this fact. In Example 2.1 we showed that the Fourier series of $f(x) = x$ is

$$x \sim \sum_{n=1}^{\infty} \frac{2(-1)^{n+1}}{n} \sin nx.$$

If we differentiate this series term by term then we obtain the series

$$\sum_{n=1}^{\infty} 2(-1)^{n+1} \cos nx.$$

This series is not the Fourier series of $f'(x) = 1$, since the Fourier series of $f'(x) = 1$ is simply the function 1. (In fact this series is not a Fourier series of any $g \in E$. The coefficients do not tend to zero, in contradiction to the Riemann-Lebesgue Lemma.)

In proving Theorem 2.15 we showed that the answer to the above question is yes if, in addition, f is continuous on all of $[-\pi, \pi]$, and $f(-\pi) = f(\pi)$. (Look again at the proof thereof, it was just integration by parts.) We formulate this result in the next theorem.

Theorem 2.26: *Assume f is continuous on $[-\pi, \pi]$, $f(-\pi) = f(\pi)$, and $f' \in E$. If*

$$f(x) \sim \frac{a_0}{2} + \sum_{n=1}^{\infty} [a_n \cos nx + b_n \sin nx]$$

is the Fourier series of f, then the Fourier series of f' is given by

$$f'(x) \sim \sum_{n=1}^{\infty}[-na_n \sin nx + nb_n \cos nx].$$

We now turn to the analogous question of when we may integrate term by term the Fourier series, and what we then get. When we differentiate the functions $\sin nx$ and $\cos nx$ we obtain $n \cos nx$ and $-n \sin nx$, respectively. The coefficient n which appears in these two expressions substantially increases the size of the coefficients of the new series. This being so, it may well cause the new series to diverge. When, on the other hand, we integrate term by term the functions $\sin nx$ and $\cos nx$ we obtain $-\frac{\cos nx}{n}$ and $\frac{\sin nx}{n}$, respectively. Thus the coefficients of the new series are substantially reduced in size. This being so, we can expect this new series to necessarily converge. And so it does. The problem which arises when integrating the Fourier series term by term has to do with the constant term $\frac{a_0}{2}$. Its integral $\frac{a_0}{2}x$ is not a term of a Fourier series. This being so, integration term by term of a Fourier series does not in general lead to a Fourier series. The main result is the following.

Theorem 2.27: *Let $f \in E$ have the Fourier series*

$$f(x) \sim \frac{a_0}{2} + \sum_{n=1}^{\infty}[a_n \cos nx + b_n \sin nx].$$

Then for each $x \in [-\pi, \pi]$,

$$\int_{-\pi}^{x} f(t)dt = \frac{a_0(x+\pi)}{2} + \sum_{n=1}^{\infty}\left[\frac{a_n}{n}\sin nx - \frac{b_n}{n}(\cos nx - \cos n\pi)\right]$$

and the series on the right-hand side converges uniformly to the function on the left.

Proof: Let us define
$$g(x) = \int_{-\pi}^{x} f(t)dt - \frac{a_0}{2}x.$$

Since $f \in E$ the function g is continuous. In addition

(2.12) $$g'(x) = f(x) - \frac{a_0}{2}$$

at each point of continuity of f. This being so, $g' \in E$. From the definition of g we have $g(-\pi) = \frac{a_0\pi}{2}$. Moreover

$$g(\pi) = \int_{-\pi}^{\pi} f(t)dt - \frac{a_0\pi}{2} = a_0\pi - \frac{a_0\pi}{2} = \frac{a_0\pi}{2}$$

and thus $g(-\pi) = g(\pi)$. The function g satisfies the conditions of Theorem 2.15. Accordingly the Fourier series of g converges uniformly to g on the interval $[-\pi, \pi]$. In particular we have equality (convergence) at each point $x \in [-\pi, \pi]$ in

(2.13) $$g(x) = \frac{A_0}{2} + \sum_{n=1}^{\infty}[A_n \cos nx + B_n \sin nx]$$

where the right-hand side is the Fourier series of g. From Theorem 2.26 (g satisfies the conditions thereof)

$$g'(x) \sim \sum_{n=1}^{\infty}[-nA_n \sin nx + nB_n \cos nx].$$

Now

$$g'(x) = f(x) - \frac{a_0}{2} \sim \sum_{n=1}^{\infty}[a_n \cos nx + b_n \sin nx].$$

Thus from (2.13) and Theorem 2.26 we have, for each $n \geq 1$, the equalities

$$nB_n = a_n, \qquad -nA_n = b_n.$$

Substituting in (2.13) we obtain

(2.14) $$\int_{-\pi}^{x} f(t)dt = g(x) + \frac{a_0}{2}x$$

$$= \frac{a_0}{2}x + \frac{A_0}{2} + \sum_{n=1}^{\infty}\left[\frac{a_n}{n}\sin nx - \frac{b_n}{n}\cos nx\right].$$

It remains to determine the coefficient A_0. To this end set $x = -\pi$ in the above to obtain

$$0 = -\frac{a_0}{2}\pi + \frac{A_0}{2} + \sum_{n=1}^{\infty} -\frac{b_n}{n}\cos n\pi.$$

Solving for A_0 gives

$$A_0 = a_0\pi + \sum_{n=1}^{\infty}\frac{2b_n}{n}\cos n\pi.$$

Substituting this value in (2.14) proves the theorem. ∎

Remark: We may, in Theorem 2.27, turn the series on the right-hand side into a Fourier series by substituting for x the Fourier series of x. However, we then lose the property of the uniform convergence of the series as the Fourier series of x does not converge uniformly to x in any neighbourhood of the points $x = \pm\pi$.

Chapter 2: Fourier Series

Exercises

1. Let $f(x) = |x|$, $-\pi \leq x \leq \pi$, and let
$$f(x) \sim \frac{a_0}{2} + \sum_{n=1}^{\infty}[a_n \cos nx + b_n \sin nx]$$
denote the Fourier series of f.
   (a) Determine the coefficients a_n and b_n.
   (b) Prove that the series $\sum_{n=1}^{\infty} na_n \sin nx$ converges for every x.
   (c) For each real x, we set $g(x) = -\sum_{n=1}^{\infty} na_n \sin nx$. Sketch the graph of g on the interval $[-2\pi, 2\pi]$.
   (d) Calculate $\sum_{n=1}^{\infty} \frac{1}{(2n-1)^2}$ and $\sum_{n=1}^{\infty} \frac{1}{(2n-1)^4}$.

2. Let $f \in E$ be an even function satisfying $\int_{-\pi}^{\pi} f(t)\,dt = 5$. Define the function F by
$$F(x) = \int_{-\pi}^{x} f(t)\,dt, \quad -\pi \leq x \leq \pi.$$
Let
$$F(x) \sim \frac{A_0}{2} + \sum_{n=1}^{\infty}[A_n \cos nx + B_n \sin nx]$$
denote the Fourier series of F and set
$$G(x) = \frac{A_0}{2} + \sum_{n=1}^{\infty}[A_n \cos nx + B_n \sin nx].$$
Calculate $G(-\pi)$, $G(\pi)$, and $G(0)$.

3. Suppose $f(x) = \begin{cases} \frac{\pi}{4}, & -\pi < x < 0, \\ \frac{\pi}{4} - x, & 0 < x < \pi. \end{cases}$
Let
$$f(x) \sim \frac{a_0}{2} + \sum_{n=1}^{\infty}[a_n \cos nx + b_n \sin nx]$$
be the Fourier series of f.
   (a) Calculate the coefficients a_n and b_n.
   (b) Set $S(x) = \frac{a_0 x}{2} + \sum_{n=1}^{\infty} \frac{1}{n}(a_n \sin nx - b_n \cos nx)$. Determine $S(x)$ for each $x \in (-\pi, \pi)$.

4. Let f be a 2π-periodic piecewise continuous function satisfying
$$\int_{-\pi}^{\pi} f(x)\,dx = 0.$$
Set $g(x) = \int_0^x f(t)\,dt$.

(a) Prove that g is 2π-periodic.

(b) Let $\sum_{n=-\infty}^{\infty} c_n e^{inx}$ be the complex Fourier series of the function f and $\sum_{n=-\infty}^{\infty} d_n e^{inx}$ the complex Fourier series of the function g. Prove that for all real x we have the equality

$$g(x) = \sum_{n=-\infty}^{\infty} d_n e^{inx}$$

where $d_n = \frac{c_n}{in}$ for every integer $n \neq 0$.

5. Let $f \in E$ and

$$f(x) \sim \frac{a_0}{2} + \sum_{n=1}^{\infty} [a_n \cos nx + b_n \sin nx]$$

be the Fourier series of f. For each $c \in [-\pi, \pi]$ prove that for all $x \in [-\pi, \pi]$ we have

$$\int_c^x f(t)\, dt = \frac{1}{2}a_0(x-c) + \sum_{n=1}^{\infty} \left[\frac{a_n}{n}(\sin nx - \sin nc) - \frac{b_n}{n}(\cos nx - \cos nc) \right]$$

where the series on the right side converges uniformly to the function on the left.

6. (a) Let $g \in E[0, \pi]$ and

$$g(x) \sim \frac{a_0}{2} + \sum_{n=1}^{\infty} a_n \cos nx$$

denote the cosine series of g. Prove that for each $x \in [0, \pi]$ we have

$$\int_0^x g(t)\, dt = \frac{a_0 x}{2} + \sum_{n=1}^{\infty} \frac{a_n}{n} \sin nx,$$

and in addition the series on the right-hand side converges uniformly to the function on the left.

(b) Let $h \in E[0, \pi]$ and

$$h(x) \sim \sum_{n=1}^{\infty} b_n \sin nx$$

denote the sine series of h. Prove that for each $x \in [0, \pi]$ we have

$$\int_0^x h(t)\, dt = \beta - \sum_{n=1}^{\infty} \frac{b_n}{n} \cos nx$$

where $\beta = \sum_{n=1}^{\infty} \frac{b_n}{n}$ and in addition the series on the right side converges uniformly to the function on the left.

Chapter 2: Fourier Series

7. Assume f is continuous on $[-\pi, \pi]$, $f(-\pi) = f(\pi)$, and $f' \in E$. Let

$$f(x) \sim \frac{a_0}{2} + \sum_{n=1}^{\infty} [a_n \cos nx + b_n \sin nx]$$

denote the Fourier series of f. Prove that

$$\lim_{n \to \infty} na_n = \lim_{n \to \infty} nb_n = 0.$$

Is it necessarily true that $\lim_{n \to \infty} n^2 a_n = \lim_{n \to \infty} n^2 b_n = 0$?

10. Fourier Series on Other Intervals

Until now we have considered Fourier series on the interval $[-\pi, \pi]$. This was mainly for convenience. Naturally the theory presented is not restricted to this particular interval. It may be extended, if done properly, to any finite interval $[a, b]$. In this section we define the Fourier series on an arbitrary closed interval $[a, b]$. We explain why the theorems and theory presented in the previous sections remain valid, in suitable form.

For each closed interval $[a, b]$, let $E[a, b]$ be the linear space of piecewise continuous functions $f : [a, b] \to \mathbb{C}$.

Definition 2.28: *Let $f \in E[a, b]$. The series*

$$\frac{a_0}{2} + \sum_{n=1}^{\infty} \left[a_n \cos \frac{2n\pi x}{b - a} + b_n \sin \frac{2n\pi x}{b - a} \right]$$

where

$$a_n = \frac{2}{b - a} \int_a^b f(x) \cos \frac{2n\pi x}{b - a} \, dx, \qquad n = 0, 1, 2, \ldots,$$

$$b_n = \frac{2}{b - a} \int_a^b f(x) \sin \frac{2n\pi x}{b - a} \, dx, \qquad n = 1, 2, \ldots,$$

is called the Fourier series of f on the interval $[a, b]$.

From the definition we see that if $[a, b] = [-\pi, \pi]$ then we obtain the usual Fourier series. We may justify this definition in a manner totally analogous to that given in the original definition of the Fourier series in Section 1 of this chapter. That is, we can prove that the sequence of functions

$$\left\{ \frac{1}{\sqrt{2}}, \ \cos \frac{2n\pi x}{b - a}, \ \sin \frac{2n\pi x}{b - a} \right\}_{n=1}^{\infty}$$

is an orthonormal system of functions in the space $E[a, b]$ with respect to the inner product
$$\langle f, g \rangle = \frac{2}{b-a} \int_a^b f(x)\overline{g(x)}\,dx.$$
However, we will justify (or more precisely, explain) Definition 2.28 in another way. Since we have already presented the general theory of Fourier series on the interval $[-\pi, \pi]$, we will use those results to ease our labours.

Let us first consider intervals of the form $[-c, c]$, where $c > 0$. Let $f \in E[-c, c]$ and
$$g(t) = f(\tfrac{ct}{\pi}), \qquad -\pi \leq t \leq \pi.$$
That is, we simply make the variable substitution $x = \tfrac{ct}{\pi}$. Obviously g is piecewise continuous in $[-\pi, \pi]$, and thus has the standard Fourier series
$$g(t) \sim \frac{a_0}{2} + \sum_{n=1}^{\infty} [a_n \cos nt + b_n \sin nt]$$
where
$$a_n = \frac{1}{\pi} \int_{-\pi}^{\pi} g(t) \cos nt\,dt, \qquad n = 0, 1, 2, \ldots,$$
$$b_n = \frac{1}{\pi} \int_{-\pi}^{\pi} g(t) \sin nt\,dt, \qquad n = 1, 2, \ldots.$$
Substituting back $t = \tfrac{\pi x}{c}$, we obtain

(2.15) $$f(x) \sim \frac{a_0}{2} + \sum_{n=1}^{\infty} \left[a_n \cos \frac{n\pi x}{c} + b_n \sin \frac{n\pi x}{c} \right]$$

where

(2.16) $$\begin{cases} a_n = \dfrac{1}{c} \int_{-c}^{c} f(x) \cos \dfrac{n\pi x}{c}\,dx, & n = 0, 1, 2, \ldots, \\[2mm] b_n = \dfrac{1}{c} \int_{-c}^{c} f(x) \sin \dfrac{n\pi x}{c}\,dx, & n = 1, 2, \ldots. \end{cases}$$

This is precisely the form given in Definition 2.28 for intervals $[a, b] = [-c, c]$.

As our next step we consider the general interval $[a, b]$. Let $f \in E[a, b]$, and define $g : \mathbb{R} \to \mathbb{C}$ as the $(b-a)$-periodic extension of f to all of $\mathbb{R}$. That is, g is the unique function defined on all of $\mathbb{R}$ satisfying the two conditions:

(a) $g(x) = f(x)$ for $x \in [a, b)$.
(b) $g(x + b - a) = g(x)$ for all $x \in \mathbb{R}$.

It follows that g is piecewise continuous on every finite interval. In particular,

Chapter 2: Fourier Series

we may define its Fourier series on the interval $[-c, c]$, where $c = \frac{b-a}{2}$. From formulae (2.15) and (2.16) we obtain

(2.17) $$g(x) \sim \frac{a_0}{2} + \sum_{n=1}^{\infty} \left[a_n \cos \frac{2n\pi x}{b-a} + b_n \sin \frac{2n\pi x}{b-a} \right]$$

where

$$a_n = \frac{2}{b-a} \int_{-c}^{c} g(x) \cos \frac{2n\pi x}{b-a} \, dx, \qquad n = 0, 1, 2, \ldots,$$

$$b_n = \frac{2}{b-a} \int_{-c}^{c} g(x) \sin \frac{2n\pi x}{b-a} \, dx, \qquad n = 1, 2, \ldots.$$

The function g and the Fourier series just defined are both $(b-a)$-periodic on all of $\mathbb{R}$. Thus (2.17) may be considered as being valid not only on $[-c, c]$, but on all of $\mathbb{R}$. In particular, formula (2.17) is valid on the interval $[a, b]$ (where $f(x) = g(x)$). In addition, the functions $g(x) \cos \frac{2n\pi x}{b-a}$ and $g(x) \sin \frac{2n\pi x}{b-a}$ are $(b-a)$-periodic on all of $\mathbb{R}$. Thus they have the same integrals on every interval of length $b-a$. (Recall that if h is periodic with period L, then the integral $\int_r^{r+L} h(x) \, dx$ is a constant independent of r.) Thus we may write

$$a_n = \frac{2}{b-a} \int_a^b f(x) \cos \frac{2n\pi x}{b-a} \, dx, \qquad n = 0, 1, 2, \ldots,$$

$$b_n = \frac{2}{b-a} \int_a^b f(x) \sin \frac{2n\pi x}{b-a} \, dx, \qquad n = 1, 2, \ldots.$$

In this way we arrive at Definition 2.28.

Example 2.12: If f is piecewise continuous on $[0, \pi]$, then its Fourier series thereon is

$$f(x) \sim \frac{a_0}{2} + \sum_{n=1}^{\infty} [a_n \cos 2nx + b_n \sin 2nx]$$

where

$$a_n = \frac{2}{\pi} \int_0^{\pi} f(x) \cos 2nx \, dx, \qquad n = 0, 1, 2, \ldots,$$

$$b_n = \frac{2}{\pi} \int_0^{\pi} f(x) \sin 2nx \, dx, \qquad n = 1, 2, \ldots.$$

We should emphasize that there is no relationship between this series and the sine or cosine series of f on $[0, \pi]$!

Finally, the complex Fourier series of a piecewise continuous function $f : [a, b] \to \mathbb{C}$ is

$$f(x) \sim \sum_{n=-\infty}^{\infty} c_n e^{\frac{2in\pi x}{b-a}}$$

where
$$c_n = \frac{1}{b-a} \int_a^b f(x) e^{-\frac{2\pi i n x}{b-a}} \, dx, \qquad n = 0, \pm 1, \pm 2, \ldots.$$

Exercises

1. Let $f(x) = x^2$ and
$$\frac{a_0}{2} + \sum_{n=1}^{\infty} [a_n \cos nx + b_n \sin nx]$$
be the Fourier series of f on $[-\pi, \pi]$. Let
$$\frac{A_0}{2} + \sum_{n=1}^{\infty} [A_n \cos nx + B_n \sin nx]$$
be the Fourier series of f on $[0, 2\pi]$. Define the function
$$h(x) = \frac{a_0 - A_0}{2} + \sum_{n=1}^{\infty} (a_n - A_n) \cos nx + (b_n - B_n) \sin nx.$$
Calculate h and give an exact sketch of the graph of h on $[-\pi, 2\pi]$.

2. Let $f : \mathbb{R} \to \mathbb{C}$ be a piecewise continuous function which is π-periodic. Let
$$f(x) \sim \frac{a_0}{2} + \sum_{n=1}^{\infty} [a_n \cos 2nx + b_n \sin 2nx]$$
be the Fourier series of f on $[0, \pi]$, and
$$f(x) \sim \frac{A_0}{2} + \sum_{n=1}^{\infty} [A_n \cos nx + B_n \sin nx]$$
the Fourier series of f on $[-\pi, \pi]$. Express the A_n and B_n in terms of the a_n and b_n.

3. Let
$$f(x) = \begin{cases} A \sin \omega_0 t, & 0 < t < \frac{T}{2}, \\ 0, & \frac{T}{2} \le t < T, \end{cases}$$
where $\omega_0 = \frac{2\pi}{T}$. Calculate the complex Fourier series of f on the interval $[0, T]$.

4. Let $f \in E[a, b]$. Determine Parseval's identity associated with the Fourier series of f on $[a, b]$, and explain why it is correct.

5. Let $f \in E[-\pi, \pi]$ and

$$f(x) \sim \frac{a_0}{2} + \sum_{n=1}^{\infty}[a_n \cos nx + b_n \sin nx]$$

be the Fourier series of f. Assume that f is $\frac{\pi}{m}$-periodic, for some $m \in \mathbb{N}$. Prove that $a_n = b_n = 0$ for every n which is not divisible by $2m$.

11. Applications to Partial Differential Equations

This section is intended mainly for the reader with some background or interest in the subject of partial differential equations. Fourier series were originally developed (by Fourier himself) as a method and tool to be used in solving certain problems in the area of partial differential equations. In this section we present one of the simpler methods of solving partial differential equations by the use of Fourier series. This method is called *separation of variables* (or sometimes the *Fourier method*). We demonstrate this method by considering the homogeneous heat equation defined on a rod of length $2L$ with periodic boundary conditions. In mathematical terms we must find a solution $u = u(x, t)$ to the problem

(2.18)
$$\begin{cases} u_t - ku_{xx} = 0, & -L < x < L, \quad 0 < t < \infty, \\ u(x, 0) = f(x), & -L \le x \le L, \\ u(-L, t) = u(L, t), & 0 \le t < \infty, \\ u_x(-L, t) = u_x(L, t), & 0 \le t < \infty, \end{cases}$$

where $k > 0$ is a constant (which has a certain physical meaning). The common wisdom is that these mathematical equations model (under ideal conditions) the heat flow ($u(x, t)$ is the temperature) in a ring of length $2L$, where the initial ($t = 0$) distribution of temperature in the ring is given by the function f. A point (or cross-section) on the ring is represented by a point in the interval $[-L, L]$, where the endpoints $x = L$ and $x = -L$ represent one and the same point on the ring. For this reason the mathematical representation of this problem includes the equations $u(-L, t) = u(L, t)$ and $u_x(-L, t) = u_x(L, t)$. To obtain a "good" solution to this problem, it is better if we assume that f is a continuous function, $f' \in E$, and f satisfies $f(-L) = f(L)$ and $f'(-L) = f'(L)$.

The idea behind the method of separation of variables is to first find all non-trivial (not identically zero) solutions of the form $u(x, t) = X(x)T(t)$ to the

11. Applications to Partial Differential Equations

homogeneous system

(2.19)
$$\begin{cases} u_t - ku_{xx} = 0, & -L < x < L, \quad 0 < t < \infty, \\ u(-L, t) = u(L, t), & 0 \le t < \infty, \\ u_x(-L, t) = u_x(L, t), & 0 \le t < \infty. \end{cases}$$

Later we will look for a solution to the equation $u(x, 0) = f(x)$ from the linear space generated by these solutions of (2.19). Consider (2.19) and $u(x, t) = X(x)T(t)$. Then

$$u_t(x, t) = X(x)T'(t), \quad u_{xx}(x, t) = X''(x)T(t).$$

Substituting these forms in the equation we obtain

$$X(x)T'(t) - kX''(x)T(t) = 0$$

and thus

$$X(x)T'(t) = kX''(x)T(t).$$

Dividing both sides of the equation by $kX(x)T(t)$ (assuming we are not dividing by zero) we obtain

$$\frac{T'(t)}{kT(t)} = \frac{X''(x)}{X(x)}.$$

The expression on the left-hand side is a function of t alone, while the expression on the right-hand side is a function only of x. Since the variables x and t do not depend upon each other, the above equality can hold if and only if both sides of the equation are equal to some (unknown) constant $-\lambda$ for all values of x and t. Thus we may write

$$\frac{T'(t)}{kT(t)} = \frac{X''(x)}{X(x)} = -\lambda.$$

This being so, we obtain the pair of ordinary differential equations with unknown constant λ:

$$X''(x) + \lambda X(x) = 0,$$

$$T'(t) + k\lambda T(t) = 0.$$

We have, in addition, the two boundary conditions from which we will derive two conditions. From the boundary condition $u(-L, t) = u(L, t)$ it follows that for all $t \ge 0$,

$$X(-L)T(t) = X(L)T(t).$$

Chapter 2: Fourier Series

There exist two possibilities. Either $T(t) = 0$ for all $t \geq 0$, or $X(-L) = X(L)$. The first possibility leads us to the trivial solution which does not interest us. Thus we are left with the condition $X(-L) = X(L)$. In the same way we obtain the condition $X'(-L) = X'(L)$ from the remaining boundary condition. Thus in our search for all (non-trivial) solutions to (2.19) of the form $u(x,t) = X(x)T(t)$ we are led to the equations for X:

(2.20) $$\begin{cases} X''(x) + \lambda X(x) = 0, & 0 < x < L, \\ X(-L) = X(L), \\ X'(-L) = X'(L). \end{cases}$$

At this stage we find all possible values λ for which there is a non-trivial solution to (2.20). It may be checked that the values of λ for which there exist non-trivial solutions to (2.20) are exactly

$$\lambda_n = \frac{n^2 \pi^2}{L^2}, \qquad n = 0, 1, 2, \ldots.$$

For $\lambda_0 = 0$ the equation is $X''(x) = 0$ and the general solution is

$$X(x) = c_1 x + c_2.$$

From the condition $X(-L) = X(L)$ we obtain $c_1 = 0$, while the condition $X'(-L) = X'(L)$ is always satisfied. This being so, in this case, the constant functions $X(x) = C$ are the solutions to (2.20).

For $\lambda_n = \frac{n^2\pi^2}{L^2}$, $n \geq 1$, the equation is

$$X''(x) + \frac{n^2 \pi^2}{L^2} X(x) = 0$$

and the general solution has the form

$$X(x) = c_1 \sin \frac{n\pi}{L} x + c_2 \cos \frac{n\pi}{L} x.$$

The boundary conditions $X(-L) = X(L)$ and $X'(-L) = X'(L)$ are satisfied by each and every one of these solutions.

In summary we can say that for all $n \in \mathbb{N}$ and $\lambda_n = \frac{n^2\pi^2}{L^2}$ we have two non-trivial linearly independent solutions

$$X_n(x) = \cos \frac{n\pi x}{L}, \qquad X_n^*(x) = \sin \frac{n\pi x}{L}.$$

Every other solution is a linear combination of these two solutions. The values λ_n are called the *eigenvalues* of the problem, and the solutions X_n and X_n^* are called the *eigenfunctions* associated with the eigenvalue λ_n. We also recall that among the eigenvalues we also have $\lambda_0 = 0$, with associated eigenfunction

$$X_0(x) = 1.$$

We now consider the second equation $T'(t) + k\lambda T(t) = 0$. We naturally restrict ourselves to $\lambda = \lambda_n = \frac{n^2\pi^2}{L^2}$, $n = 0, 1, 2, 3, \ldots$. For each such n there is the non-trivial solution

$$T_n(t) = e^{-k\lambda_n t}.$$

Every other solution is a constant multiple thereof. In summary, for each $n \in \mathbb{N}$ we have a pair of non-trivial solutions to (2.19) of the form

$$u_n(x,t) = X_n(x)T_n(t) = e^{-k\lambda_n t} \cos \frac{n\pi x}{L},$$

$$u_n^*(x,t) = X_n^*(x)T_n(t) = e^{-k\lambda_n t} \sin \frac{n\pi x}{L}.$$

For $n = 0$ we have the solution

$$u_0(x,t) = X_0(t)T_0(t) = 1.$$

Since the system (2.19) is homogeneous every "infinite linear combination" of the solutions is again a solution (assuming it converges). So we have, in a sense, an infinity of solutions of the general form

$$(2.21) \qquad u(x,t) = \frac{a_0}{2} + \sum_{n=1}^{\infty} e^{-k\lambda_n t}\left[a_n \cos\frac{n\pi x}{L} + b_n \sin\frac{n\pi x}{L}\right].$$

One can in fact prove that this is the general form of the solution of (2.19). Next we turn to the problem of finding a solution to (2.18). We must consider the non-homogeneous initial condition $u(x,0) = f(x)$, $-L \leq x \leq L$. This condition should determine the two sequences of coefficients $\{a_n\}_{n=0}^{\infty}$ and $\{b_n\}_{n=1}^{\infty}$. If we substitute $t = 0$ in the general solution we obtain

$$f(x) = u(x,0) = \frac{a_0}{2} + \sum_{n=1}^{\infty}\left[a_n \cos\frac{n\pi x}{L} + b_n \sin\frac{n\pi x}{L}\right].$$

We recognize this as the Fourier series of f on the interval $[-L, L]$. We know how to find the requisite a_n and b_n. From (2.16) in the previous section we have

$$a_n = \frac{1}{L}\int_{-L}^{L} f(x)\cos\frac{n\pi x}{L}\,dx, \qquad n = 0, 1, 2, \ldots,$$

$$b_n = \frac{1}{L}\int_{-L}^{L} f(x)\sin\frac{n\pi x}{L}\,dx, \qquad n = 1, 2, \ldots,$$

and this "solves" (2.18). It may be proven that with these coefficients the series (2.21) (and also its partial derivatives of every order) converges uniformly and absolutely for every value of $x \in [-L, L]$ and $t \in (0, \infty)$. We may also verify the fact that the above series satisfies the heat equation.

Review Exercises

1. Let $f(x) = x + \cos x$ and
$$f(x) \sim \frac{a_0}{2} + \sum_{n=1}^{\infty}\left[a_n \cos \frac{nx}{2} + b_n \sin \frac{nx}{2}\right]$$
be the Fourier series of f on $[0, 4\pi]$.
   (a) Determine the a_n and b_n.
   (b) Let $g(x) = \frac{A_0}{2} + \sum_{n=1}^{5} B_n \sin \frac{nx}{2}$. For what values of A_0 and B_n, $1 \leq n \leq 5$, is the distance between f and g minimal?

2. Let $f(x) = x^2$ and
$$f(x) \sim \frac{a_0}{2} + \sum_{n=1}^{\infty}[a_n \cos nx + b_n \sin nx]$$
denote the Fourier series of f on $[\pi, 3\pi]$.
   (a) Calculate the a_n and b_n.
   (b) Set $g(x) = \frac{a_0}{2} + \sum_{n=1}^{\infty} a_n \cos \frac{nx}{2}$, $-\pi \leq x \leq \pi$. Determine g and sketch the graph of g on $[-\pi, \pi]$.
   (c) Set $h(x) = \sum_{n=1}^{\infty} b_n \sin \frac{nx}{2}$, $-\pi \leq x \leq \pi$. Determine h and sketch the graph of h on $[-\pi, \pi]$.

3. Set $f(x) = |\sin x|$ and let
$$f(x) \sim \frac{a_0}{2} + \sum_{n=1}^{\infty}[a_n \cos nx + b_n \sin nx]$$
denote the Fourier series of f on $[-\pi, \pi]$.
   (a) Calculate the a_n and b_n.
   (b) Set $g(x) = \sum_{n=1}^{\infty}[-na_n \sin nx + nb_n \cos nx]$. Determine g and sketch the graph of g on $[-\pi, \pi]$.
   (c) Calculate the sums
$$\sum_{n=1}^{\infty} \frac{1}{4n^2 - 1}, \quad \sum_{n=1}^{\infty} \frac{(-1)^n}{4n^2 - 1}, \quad \sum_{n=1}^{\infty} \frac{1}{(4n^2 - 1)^2}, \quad \sum_{n=1}^{\infty} \frac{n^2}{(4n^2 - 1)^2}.$$

4. Find the complex Fourier series of $f(t) = \frac{1}{1 - \frac{1}{2}e^{-it}}$ on the interval $[-\pi, \pi]$.

5. Let $f : \mathbb{R} \to \mathbb{C}$ be a 2π-periodic piecewise continuous function and
$$f(x) \sim \sum_{n=-\infty}^{\infty} c_n e^{inx}$$

denote the Fourier series of f on $[-\pi, \pi]$. Define the function
$$g(x) = \frac{f(x) + f(x+\pi)}{2}$$
and let
$$g(x) \sim \sum_{n=-\infty}^{\infty} d_n e^{inx}$$
be the complex Fourier series of g on $[-\pi, \pi]$.

(a) Determine the coefficients d_n in terms of the coefficients c_n.
(b) For $f(x) = x$, $x \in [-\pi, \pi]$, calculate the sum $\sum_{n=1}^{\infty} |c_{2n}|^2$.

6. Let f be a π-periodic function for which
$$f(x) = \begin{cases} \sin 2x, & 0 \le x \le \frac{\pi}{2}, \\ 0, & \frac{\pi}{2} \le x \le \pi. \end{cases}$$

(a) Prove that for all $x \in \mathbb{R}$
$$f(x) = \frac{1}{\pi} + \frac{1}{2} \sin 2x - \frac{2}{\pi} \sum_{n=1}^{\infty} \frac{\cos 4nx}{(2n-1)(2n+1)}$$
and then prove that
$$\sum_{n=1}^{\infty} \frac{1}{(2n-1)^2(2n+1)^2} = \frac{\pi^2 - 8}{16}.$$

(b) Determine the value of the sum
$$\frac{\sin 4x}{1 \cdot 2 \cdot 3} + \frac{\sin 8x}{3 \cdot 4 \cdot 5} + \frac{\sin 12x}{5 \cdot 6 \cdot 7} + \cdots, \quad 0 \le x \le \pi.$$

7. Let
$$f(x) = \begin{cases} 0, & -\pi \le x < 0, \\ e^x, & 0 \le x \le \pi, \end{cases}$$
and set
$$F(x) = \frac{a_0}{2} + \sum_{n=1}^{\infty} [a_n \cos nx + b_n \sin nx]$$
where $\frac{a_0}{2} + \sum_{n=1}^{\infty} [a_n \cos nx + b_n \sin nx]$ is the Fourier series of f on $[-\pi, \pi]$.

(a) Calculate the a_n and b_n.
(b) Determine the values $F(0)$, $F(\frac{\pi}{2})$, and $F(\pi)$.
(c) Sketch the graph of F on the interval $[-2\pi, 2\pi]$.
(d) Calculate the sums $\sum_{n=1}^{\infty} \frac{1}{n^2+1}$ and $\sum_{n=1}^{\infty} \frac{(-1)^n}{n^2+1}$.

Chapter 2: Fourier Series

8. Let $f(x) = e^{x(1+2\pi i)}$.
   (a) Determine the complex Fourier series of f on the interval $[-1, 1]$.
   (b) Determine the usual Fourier series of f on the interval $[-1, 1]$.
   (c) Calculate the sums $\sum_{n=1}^{\infty} \frac{1}{1+n^2\pi^2}$ and $\sum_{n=1}^{\infty} \frac{(-1)^n}{1+n^2\pi^2}$.

9. (a) Determine the Fourier series of the function
$$f(x) = \begin{cases} 0, & -\pi < x < 0, \\ \sin x, & 0 < x < \pi, \end{cases}$$
on the interval $[-\pi, \pi]$.

   (b) Using (a) prove that
$$\frac{\pi - 2}{4} = \frac{1}{1 \cdot 3} - \frac{1}{3 \cdot 5} + \frac{1}{5 \cdot 7} - \frac{1}{7 \cdot 9} + \cdots.$$

   (c) Show that $\dfrac{1}{2} = \sum_{n=1}^{\infty} \dfrac{1}{4n^2 - 1}$.

10. We define the function
$$f(x) = \begin{cases} \sin 2x, & -\frac{\pi}{2} \leq x \leq \frac{\pi}{2}, \\ 0, & \text{otherwise}, \end{cases}$$
on the interval $[-\pi, \pi]$.

   (a) Determine the Fourier series of f on $[-\pi, \pi]$.
   (b) Determine the Fourier series of f' on $[-\pi, \pi]$.
   (c) To what values does the Fourier series of f' converge at the points $x = \pm\frac{\pi}{2}$?
   (d) Calculate the sums
$$\sum_{k=1}^{\infty} \frac{1}{(2k-3)^2(2k+1)^2}, \quad \sum_{k=1}^{\infty} \frac{(2k-1)(-1)^k}{(2k-3)(2k+1)}.$$

11. Let f be a 2π-periodic piecewise continuous function and
$$f(x) \sim \sum_{n=1}^{\infty} [a_n \cos nx + b_n \sin nx]$$
its Fourier series on $[-\pi, \pi]$. Set
$$g(x) = \int_{-\pi}^{x} [f(t) + f(\pi - t)]\, dt$$

and let
$$g(x) \sim \frac{A_0}{2} + \sum_{n=1}^{\infty}[A_n \cos nx + B_n \sin nx]$$
be the Fourier series of g on $[-\pi, \pi]$. Express the A_n and B_n in terms of a_n and b_n.

12. Let $C[-\pi, \pi]^n$ denote the space of complex-valued continuous functions defined on the cube $[-\pi, \pi]^n$. On this linear space we define the inner product
$$\langle f, g \rangle = \frac{1}{(2\pi)^n} \underbrace{\int_{-\pi}^{\pi} \cdots \int_{-\pi}^{\pi}}_{n \text{ times}} f(x_1, \ldots, x_n) \overline{g(x_1, \ldots, x_n)} \, dx_1 \cdots dx_n.$$

Prove that the family of functions
$$\left\{ e^{i(m_1 x_1 + \cdots + m_n x_n)} \right\}_{m_1, \ldots, m_n \in \mathbb{Z}}$$
is an orthonormal system with respect to the above inner product. If we define
$$c_{m_1, \ldots, m_n} = \frac{1}{(2\pi)^n} \underbrace{\int_{-\pi}^{\pi} \cdots \int_{-\pi}^{\pi}}_{n \text{ times}} f(x_1, \ldots, x_n) e^{-i(m_1 x_1 + \cdots + m_n x_n)} dx_1 \cdots dx_n$$

then the series
$$\sum_{m_1, \ldots, m_n = -\infty}^{\infty} c_{m_1, \ldots, m_n} e^{i(m_1 x_1 + \cdots + m_n x_n)}$$

is called the *multivariate complex Fourier series* of f. Prove that if $f_1, f_2, \ldots, f_n$ are functions in $C[-\pi, \pi]$ and
$$f(x_1, x_2, \ldots, x_n) = f_1(x_1) f_2(x_2) \cdots f_n(x_n)$$
then
$$c_{m_1, \ldots, m_n} = a_{m_1}^1 a_{m_2}^2 \cdots a_{m_n}^n$$
where the a_m^k are the coefficients of the complex Fourier series of f_k. That is, $f_k(x) \sim \sum_{m=-\infty}^{\infty} a_m^k e^{imx}$.

Chapter 3
The Fourier Transform

0. Introduction

In this chapter we examine the Fourier transform. In the first six sections we present the basic theory of the Fourier transform. Special attention should be paid to the topics of the inverse transform, Plancherel's Theorem, and convolution. In Section 3.7 we apply the Fourier transform to the solution of certain partial differential equations. We use the Fourier transform (and Fourier series) in Section 3.8 to solve some problems from information and signal theory. The Fourier transform is the first of the integral transforms studied in this book. The name of Fourier is justifiably attached to this transform. It was Fourier who first defined and used it.

1. Definitions and Basic Properties

Let f be a function defined on all of $\mathbb{R}$ with values in $\mathbb{C}$. We formally define the function $F : \mathbb{R} \to \mathbb{C}$ by

$$F(\omega) = \mathcal{F}[f](\omega) = \frac{1}{2\pi} \int_{-\infty}^{\infty} f(x) e^{-i\omega x} \, dx.$$

This definition is formal in that the integral on the right-hand side need not exist. If it exists then the function F (or $\mathcal{F}[f]$) is called the *Fourier transform* of f. In most books the notation $\hat{f}(\omega)$ is used for the Fourier transform of f. For the reader's convenience we will use F or $\mathcal{F}[f]$. There is an analogy between Fourier series and the Fourier transform which we will not delve into here. We do, however, wish to stress that the Fourier series is intended for functions defined on a finite interval (or periodic functions defined on all of $\mathbb{R}$), while the Fourier transform is used for functions defined on all of $\mathbb{R}$ (which are not periodic).

We denote by $G(\mathbb{R})$ the family of functions defined on $\mathbb{R}$ with values in $\mathbb{C}$ which are piecewise continuous and absolutely integrable. We note that:

(a) f is *piecewise continuous* on all of $\mathbb{R}$ if it is piecewise continuous on each finite interval $[a, b]$. This being so, f might have an infinite number of discontinuities (but only a finite number in each finite subinterval). A simple example of such a function is $f(x) = [x]$, the integer value function. This function has a jump discontinuity at each integer.

(b) f is *absolutely integrable* on $\mathbb{R}$ if $\int_{-\infty}^{\infty} |f(x)|\, dx < \infty$. That is, the integral over $\mathbb{R}$ of $|f|$ exists and is finite.

It is not difficult to verify that $G(\mathbb{R})$ is a linear space over $\mathbb{C}$. From the definition it follows that for each $f \in G(\mathbb{R})$ the Fourier transform of f is defined for all $\omega \in \mathbb{R}$. The Fourier transform also has a number of other properties.

Theorem 3.1: *For each $f \in G(\mathbb{R})$,*
1. $F(\omega)$ *is defined for all $\omega \in \mathbb{R}$;*
2. $F \in C(\mathbb{R})$, *i.e., F is a continuous function on $\mathbb{R}$;*
3. $\lim_{\omega \to \pm\infty} F(\omega) = 0$.

To prove this theorem we use the *Lebesgue Dominated Convergence Theorem* which is a well-known theorem from classical analysis. However, its proof is, unfortunately, beyond the scope of this book.

Theorem 3.2: **(Lebesgue Dominated Convergence Theorem)** *Let $\{f_h\}_{h \in \mathbb{R}}$ be a family of piecewise continuous functions. Assume*

1. *There exists a function g such that $|f_h(x)| \leq g(x)$ for all $x \in \mathbb{R}$ and all $h \in \mathbb{R}$;*
2. $\int_{-\infty}^{\infty} g(x)\, dx < \infty$;
3. $\lim_{h \to 0} f_h(x) = f(x)$ *for every $x \in \mathbb{R}$.*

Then
$$\lim_{h \to 0} \int_{-\infty}^{\infty} f_h(x)\, dx = \int_{-\infty}^{\infty} f(x)\, dx.$$

One of the central problems in analysis has to do with the "infinity", and how to deal with it. One aspect of this problem is the ability or inability to interchange mathematical operations while maintaining the same values (see also Appendix B). The Lebesgue Dominated Convergence Theorem deals with one such problem. It provides conditions under which it is permissible to change the order of the actions of integration and taking limits. In general this cannot be done. Here is a simple example. Let

$$f_n(x) = \begin{cases} 1, & n \leq x \leq n+1, \\ 0, & \text{otherwise,} \end{cases}$$

and let $f \equiv 0$ (i.e., $f(x) = 0$ for all x). It is easily checked that for each $x \in \mathbb{R}$ we have $\lim_{n \to \infty} f_n(x) = f(x)$, while $\int_{-\infty}^{\infty} f_n(x)\,dx = 1$ for each $n \in \mathbb{N}$. Thus

$$\lim_{n \to \infty} \int_{-\infty}^{\infty} f_n(x)\,dx = 1 \neq 0 = \int_{-\infty}^{\infty} \lim_{n \to \infty} f_n(x)\,dx.$$

The fact that we are here dealing with an improper integral on all of $\mathbb{R}$ is not the relevant issue. This same type of phenomenon can occur when considering the finite interval. The existence of g satisfying Conditions 1 and 2 in the statement of the theorem permits us to change the order of operations and maintain the same value.

Proof of Theorem 3.1: 1. From the fact that $\left|e^{-i\omega x}\right| = 1$ for all real x and ω, it follows that

$$\frac{1}{2\pi} \int_{-\infty}^{\infty} \left|f(x)e^{-i\omega x}\right| dx = \frac{1}{2\pi} \int_{-\infty}^{\infty} |f(x)|\,dx < \infty.$$

Thus the function $f(x)e^{-i\omega x}$ is absolutely integrable on $\mathbb{R}$, for each real ω. In addition $f(x)e^{-i\omega x}$ is piecewise continuous and thus belongs to $G(\mathbb{R})$. This being so, $F(\omega)$ is well defined for every real ω.

2. Let $\omega \in \mathbb{R}$. We will prove that F is continuous at ω. In other words we will prove that

$$\lim_{h \to 0}[F(\omega + h) - F(\omega)] = 0.$$

From the definition of F we have

$$F(\omega + h) - F(\omega) = \frac{1}{2\pi} \int_{-\infty}^{\infty} f(x)e^{-i(\omega+h)x}\,dx - \frac{1}{2\pi} \int_{-\infty}^{\infty} f(x)e^{-i\omega x}\,dx$$

$$= \frac{1}{2\pi} \int_{-\infty}^{\infty} f(x)e^{-i\omega x}\left[e^{-ihx} - 1\right] dx.$$

For each real h we define the function

$$f_h(x) = f(x)e^{-i\omega x}\left[e^{-ihx} - 1\right].$$

Now for every $x \in \mathbb{R}$

$$\lim_{h \to 0} f_h(x) = \lim_{h \to 0} f(x)e^{-i\omega x}\left[e^{-ihx} - 1\right]$$

$$= f(x)e^{-i\omega x} \lim_{h \to 0}\left[e^{-ihx} - 1\right]$$

$$= f(x)e^{-i\omega x} \cdot 0$$

$$= 0.$$

In addition

$$|f_h(x)| = |f(x)| \cdot |e^{-i\omega x}| \cdot |e^{-ihx} - 1| \leq |f(x)| \cdot 1 \cdot 2 = 2|f(x)|.$$

The function $g(x) = 2|f(x)|$ satisfies Conditions 1 and 2 of Theorem 3.2. As a consequence we obtain

$$\lim_{h \to 0} \frac{1}{2\pi} \int_{-\infty}^{\infty} f_h(x)\,dx = 0.$$

In other words $\lim_{h \to 0}[F(\omega + h) - F(\omega)] = 0$, and thus F is continuous at every point $\omega \in \mathbb{R}$.

3. By definition

$$\lim_{\omega \to \pm\infty} F(\omega) = \lim_{\omega \to \pm\infty} \int_{-\infty}^{\infty} f(x) e^{-i\omega x}\,dx$$

$$= \lim_{\omega \to \pm\infty} \left[\int_{-\infty}^{\infty} f(x) \cos \omega x\,dx - i \int_{-\infty}^{\infty} f(x) \sin \omega x\,dx \right].$$

To prove that $\lim_{\omega \to \pm\infty} F(\omega) = 0$ it suffices to prove

$$\lim_{\omega \to \pm\infty} \int_{-\infty}^{\infty} f(x) \cos \omega x\,dx = 0$$

and

$$\lim_{\omega \to \pm\infty} \int_{-\infty}^{\infty} f(x) \sin \omega x\,dx = 0.$$

We prove only the first result. The proof of the second is totally analogous. (Note that this is essentially an analogue of the Riemann-Lebesgue Lemma.)

We are given an f defined on $\mathbb{R}$ which is absolutely integrable. Choose $\epsilon > 0$. There exists an $M > 0$ such that

$$\int_{|x|>M} |f(x)|\,dx < \epsilon.$$

Thus

$$\left| \int_{|x|>M} f(x) \cos \omega x\,dx \right| \leq \int_{|x|>M} |f(x) \cos \omega x|\,dx \leq \int_{|x|>M} |f(x)|\,dx < \epsilon.$$

Since the function f is piecewise continuous on the interval $[-M, M]$, there exists a partition of the interval $[-M, M]$,

$$-M = x_0 < x_1 < \cdots < x_m = M,$$

such that the step function

$$h(x) = f(x_k), \qquad x_{k-1} < x \leq x_k, \quad k = 1, 2, \ldots, m,$$

satisfies
$$\int_{-M}^{M} |f(x) - h(x)|\, dx < \epsilon.$$

(In other words, we may approximate f, in the above sense, by the step function h. This fact is a consequence of the definition of the Riemann integral.) Now

$$\int_{-M}^{M} f(x) \cos \omega x\, dx = \int_{-M}^{M} [f(x) - h(x)] \cos \omega x\, dx + \int_{-M}^{M} h(x) \cos \omega x\, dx.$$

For each $\omega \in \mathbb{R}$

$$\left| \int_{-M}^{M} [f(x) - h(x)] \cos \omega x\, dx \right| \leq \int_{-M}^{M} |f(x) - h(x)| \cdot |\cos \omega x|\, dx$$

$$\leq \int_{-M}^{M} |f(x) - h(x)|\, dx$$

$$< \epsilon.$$

We now consider the second integral (and this is the real essence of the proof).

$$\left| \int_{-M}^{M} h(x) \cos \omega x\, dx \right| = \left| \sum_{k=1}^{m} \int_{x_{k-1}}^{x_k} f(x_k) \cos \omega x\, dx \right|$$

$$= \left| \sum_{k=1}^{m} f(x_k) \frac{\sin \omega x_k - \sin \omega x_{k-1}}{\omega} \right|$$

$$\leq \sum_{k=1}^{m} |f(x_k)| \frac{2}{\omega}$$

$$\leq \frac{2m}{\omega} \max_{-M \leq x \leq M} |f(x)|.$$

Thus for ω sufficiently large

$$\frac{2m}{\omega} \max_{-M \leq x \leq M} |f(x)| < \epsilon.$$

This being so, for ω sufficiently large

$$\left| \int_{-\infty}^{\infty} f(x) \cos \omega x\, dx \right| < 3\epsilon$$

and the theorem is proved. ∎

2. Examples

Example 3.1: Let $f(x) = e^{-|x|}$. It is readily verified that f is absolutely integrable on $\mathbb{R}$. We will determine the Fourier transform of f. By properties of oddness and evenness

$$F(\omega) = \frac{1}{2\pi} \int_{-\infty}^{\infty} e^{-|x|} e^{-i\omega x} \, dx = \frac{1}{\pi} \int_{0}^{\infty} e^{-x} \cos \omega x \, dx.$$

Integrating by parts, twice, and using the fact that $\lim_{x \to \infty} e^{-x} = 0$, we obtain

$$F(\omega) = \frac{1}{\pi} \left[1 - \omega^2 \int_{0}^{\infty} e^{-x} \cos \omega x \, dx \right].$$

Solving for $\int_0^\infty e^{-x} \cos \omega x \, dx$, we have

$$F(\omega) = \frac{1}{\pi(1 + \omega^2)}.$$

The Fourier transform of $e^{-|x|}$ is $\frac{1}{\pi(1+\omega^2)}$.

Example 3.2: Set

$$f(x) = \begin{cases} 1, & a \leq x \leq b, \\ 0, & \text{otherwise}, \end{cases}$$

where $-\infty < a < b < \infty$. We now determine the Fourier transform of f.

$$F(\omega) = \frac{1}{2\pi} \int_{-\infty}^{\infty} f(x) e^{-i\omega x} \, dx = \frac{1}{2\pi} \int_{a}^{b} e^{-i\omega x} \, dx$$

$$= \frac{1}{2\pi} \frac{e^{-i\omega x}}{-i\omega} \bigg|_{a}^{b} = \frac{e^{-i\omega a} - e^{-i\omega b}}{2\pi i \omega}.$$

In the special case $a = -b$, $b > 0$, we obtain

$$F(\omega) = \frac{e^{i\omega b} - e^{-i\omega b}}{2\pi i \omega} = \frac{\sin \omega b}{\omega \pi}.$$

The function $\frac{\sin \omega}{\omega}$ is sometimes denoted as $\text{sinc}(\omega)$. This being so, we may also write

$$F(\omega) = \frac{b}{\pi} \text{sinc}(b\omega).$$

Example 3.3: We calculate the Fourier transform of $f(x) = e^{-x^2}$, $-\infty < x < \infty$, by two distinct methods.

Method 1: This is a somewhat circuitous method. By definition

$$F(\omega) = \frac{1}{2\pi} \int_{-\infty}^{\infty} e^{-x^2} e^{-i\omega x} \, dx$$

Chapter 3: The Fourier Transform

and thus
$$F'(\omega) = \frac{1}{2\pi} \frac{d}{d\omega} \int_{-\infty}^{\infty} e^{-x^2} e^{-i\omega x} \, dx.$$

(We may prove by Leibniz's rule (see Appendix B) that it is permissible to exchange the order of integration and differentiation.) This being so,

$$F'(\omega) = \frac{1}{2\pi} \int_{-\infty}^{\infty} e^{-x^2}(-ix) e^{-i\omega x} \, dx = \frac{i}{4\pi} \int_{-\infty}^{\infty} (-2x e^{-x^2}) e^{-i\omega x} \, dx.$$

Integrating by parts, we obtain

$$F'(\omega) = \frac{i}{4\pi} \left[e^{-x^2} e^{-i\omega x} \Big|_{-\infty}^{\infty} - \int_{-\infty}^{\infty} e^{-x^2} (-i\omega e^{-i\omega x}) \, dx \right].$$

The value of the first term on the right-hand side is given by

$$e^{-x^2} e^{-i\omega x} \Big|_{-\infty}^{\infty} = \lim_{L, M \to \infty} \left[e^{-L^2} e^{-i\omega L} - e^{-M^2} e^{i\omega M} \right] = 0.$$

Therefore

$$F'(\omega) = -\frac{i}{4\pi} \int_{-\infty}^{\infty} e^{-x^2} (-i\omega) e^{-i\omega x} \, dx$$

$$= -\frac{\omega}{2} \cdot \frac{1}{2\pi} \int_{-\infty}^{\infty} e^{-x^2} e^{-i\omega x} \, dx$$

$$= -\frac{\omega}{2} F(\omega).$$

We have not yet calculated F. But we have shown it to be a function which satisfies the first order ordinary differential equation

$$F'(\omega) + \frac{\omega}{2} F(\omega) = 0.$$

It is well known (and hopefully the reader can prove it) that the general solution to this equation is

$$F(\omega) = A e^{-\frac{\omega^2}{4}}$$

where A is an arbitrary constant. To properly determine F we must find A. To this end we need an initial condition for F. We substitute $\omega = 0$ in the definition of F to obtain

$$A = F(0) = \frac{1}{2\pi} \int_{-\infty}^{\infty} e^{-x^2} e^0 \, dx = \frac{1}{2\pi} \int_{-\infty}^{\infty} e^{-x^2} \, dx.$$

It is well known that

$$\int_{-\infty}^{\infty} e^{-x^2} \, dx = \sqrt{\pi}$$

(see Exercise 2 below), and hence

$$A = F(0) = \frac{1}{2\pi}\int_{-\infty}^{\infty} e^{-x^2}\,dx = \frac{1}{2\sqrt{\pi}}.$$

As a consequence

$$F(\omega) = \frac{1}{2\sqrt{\pi}}e^{-\frac{\omega^2}{4}}.$$

Method 2:

$$F(\omega) = \frac{1}{2\pi}\int_{-\infty}^{\infty} e^{-x^2}e^{-i\omega x}\,dx$$

$$= \frac{1}{2\pi}\int_{-\infty}^{\infty} e^{-(x+\frac{i\omega}{2})^2-\frac{\omega^2}{4}}\,dx$$

$$= e^{-\frac{\omega^2}{4}}\cdot\frac{1}{2\pi}\int_{-\infty}^{\infty} e^{-(x+\frac{i\omega}{2})^2}\,dx.$$

We now prove that

$$\int_{-\infty}^{\infty} e^{-(x+\frac{i\omega}{2})^2}\,dx = \sqrt{\pi},$$

independent of ω. As above, we take it as given that

$$\int_{-\infty}^{\infty} e^{-x^2}\,dx = \sqrt{\pi}.$$

Assume $\omega > 0$. For each $R > 0$, let γ_R denote the simple closed curve given by the union of the four straight lines

$$\gamma_1 = \{z \mid z = x,\ -R \leq x \leq R\},$$

$$\gamma_2 = \{z \mid z = R + iy,\ 0 \leq y \leq \tfrac{\omega}{2}\},$$

$$\gamma_3 = \{z \mid z = x + \tfrac{i\omega}{2},\ -R \leq x \leq R\},$$

$$\gamma_4 = \{z \mid z = -R + iy,\ 0 \leq y \leq \tfrac{\omega}{2}\},$$

put together so that γ_R has positive orientation.

The function e^{-z^2} is analytic on all of $\mathbb{C}$. Thus by Cauchy's Theorem (see Appendix A) for every $R > 0$

$$0 = \oint_{\gamma_R} e^{-z^2}\,dz$$

$$= \int_{\gamma_1} e^{-z^2}\,dz + \int_{\gamma_2} e^{-z^2}\,dz - \int_{\gamma_3} e^{-z^2}\,dz - \int_{\gamma_4} e^{-z^2}\,dz.$$

On γ_2 and γ_4

$$|e^{-z^2}| = e^{-R^2+y^2},\quad 0 \leq y \leq \frac{\omega}{2}.$$

Chapter 3: The Fourier Transform

Since both γ_2 and γ_4 are of fixed length $\frac{\omega}{2}$, it easily follows that

$$\lim_{R\to\infty} \int_{\gamma_k} e^{-z^2} dz = 0, \quad k = 2, 4.$$

Now

$$\lim_{R\to\infty} \int_{\gamma_1} e^{-z^2} dz = \lim_{R\to\infty} \int_{-R}^{R} e^{-x^2} dx = \sqrt{\pi}$$

and

$$\lim_{R\to\infty} \int_{\gamma_3} e^{-z^2} dz = \lim_{R\to\infty} \int_{-R}^{R} e^{-(x+\frac{i\omega}{2})^2} dx = \int_{-\infty}^{\infty} e^{-(x+\frac{i\omega}{2})^2} dx$$

since $e^{-(x+\frac{i\omega}{2})^2} \in G(\mathbb{R})$. This proves our claim for $\omega > 0$. For $\omega < 0$ totally analogous reasoning holds.

Exercises

1. Calculate the Fourier transform of each of the following functions:

   (a) $f_a(x) = \begin{cases} 1 - \frac{|x|}{a}, & |x| < a \\ 0, & |x| \geq a \end{cases}$

   (b) $f(x) = \begin{cases} 1, & 0 \leq x < 1 \\ 0, & \text{otherwise} \end{cases}$

   (c) $f(x) = \begin{cases} 1, & |x| \leq 1 \\ 2, & 1 < |x| < 2 \\ 0, & |x| \geq 2 \end{cases}$

   (d) $f(x) = \begin{cases} \sin x, & |x| \leq \pi \\ 0, & |x| > \pi \end{cases}$

   (e) $f(x) = \begin{cases} x, & |x| \leq a \\ 0, & |x| > a \end{cases}$

   (f) $f(x) = \begin{cases} x^2, & |x| \leq 1 \\ 0, & |x| > 1 \end{cases}$

   (g) $f(x) = \begin{cases} \cos x, & |x| \leq \pi \\ 0, & |x| > \pi \end{cases}$

   (h) $f(x) = \begin{cases} e^x, & x < 0 \\ -e^{-x}, & x > 0 \end{cases}$

   (i) $f(x) = |x|e^{-|x|}$

   (j) $f(x) = \begin{cases} \sin x, & |x| \leq \frac{\pi}{2} \\ 0, & |x| > \frac{\pi}{2} \end{cases}$

2. Prove that $I = \int_{-\infty}^{\infty} e^{-x^2} dx = \sqrt{\pi}$. (Hint: It is not difficult to show that $I^2 = \int_{-\infty}^{\infty} \int_{-\infty}^{\infty} e^{-(x^2+y^2)} dx\, dy$. By a change of variables to polar coordinates prove that $I^2 = \pi$.)

3. If $f \in G(\mathbb{R})$, is $\mathcal{F}[f] \in G(\mathbb{R})$? Explain.

3. Properties and Formulae

In this section we establish a number of formulae and general properties satisfied by the Fourier transform. These will facilitate our calculations. In addition, formulae and general properties provide insights into the basic concepts associated with the transform. We use, interchangeably, the notation $\mathcal{F}[f]$ or F to denote the Fourier transform of $f \in G(\mathbb{R})$.

3.1. Linearity: For each $f, g \in G(\mathbb{R})$ and every $a, b \in \mathbb{C}$, the function $af + bg$ is in $G(\mathbb{R})$ and

$$\mathcal{F}[af + bg](\omega) = a\mathcal{F}[f](\omega) + b\mathcal{F}[g](\omega)$$

This property is an immediate consequence of the linearity of the indefinite integral.

3.2. Let $f \in G(\mathbb{R})$. If f attains only real values (i.e., $f(x) \in \mathbb{R}$ for all x), then

$$F(-\omega) = \overline{F(\omega)}$$

Proof:

$$\begin{aligned} F(-\omega) &= \frac{1}{2\pi} \int_{-\infty}^{\infty} f(x) e^{-i(-\omega)x} \, dx \\ &= \frac{1}{2\pi} \int_{-\infty}^{\infty} f(x) e^{i\omega x} \, dx \\ &= \overline{\frac{1}{2\pi} \int_{-\infty}^{\infty} f(x) e^{-i\omega x} \, dx} \\ &= \overline{F(\omega)}. \end{aligned}$$ ∎

3.3. If $f \in G(\mathbb{R})$ is an even real-valued function, then F is even and real-valued. If f is an odd real-valued function, then F is odd and purely imaginary.

Proof: We only prove the first half of the claim. The second half is similarly proven. Assume f is even and real-valued. Then

$$F(\omega) = \frac{1}{2\pi} \int_{-\infty}^{\infty} f(x) e^{-i\omega x}\, dx$$

$$= \frac{1}{2\pi} \int_{-\infty}^{\infty} f(x)[\cos \omega x - i \sin \omega x]\, dx$$

$$= \frac{1}{2\pi} \int_{-\infty}^{\infty} f(x) \cos \omega x\, dx.$$

The other integral is zero since $f(x) \sin \omega x$ is an odd function of x on $\mathbb{R}$. Thus we are left only with the real part, and so F is real-valued. In addition, since $\cos \omega x$ is an even function of ω, the function

$$F(\omega) = \frac{1}{2\pi} \int_{-\infty}^{\infty} f(x) \cos \omega x\, dx.$$

is also an even function of ω. ∎

3.4. Shift Formula: Let $f \in G(\mathbb{R})$ and $a, b \in \mathbb{R}$, $a \neq 0$. The function

$$g(x) = f(ax + b)$$

belongs to $G(\mathbb{R})$, and

$$\boxed{\mathcal{F}[g](\omega) = \tfrac{1}{|a|} e^{\frac{i\omega b}{a}} \mathcal{F}[f]\left(\tfrac{\omega}{a}\right)}$$

Proof:

$$\mathcal{F}[g](\omega) = \frac{1}{2\pi} \int_{-\infty}^{\infty} f(ax+b) e^{-i\omega x}\, dx.$$

We substitute $y = ax + b$ to obtain

$$a > 0 \implies \mathcal{F}[g](\omega) = \frac{1}{2\pi} \int_{-\infty}^{\infty} f(y) e^{-i\omega\left(\frac{y-b}{a}\right)} \frac{dy}{a},$$

$$a < 0 \implies \mathcal{F}[g](\omega) = -\frac{1}{2\pi} \int_{-\infty}^{\infty} f(y) e^{-i\omega\left(\frac{y-b}{a}\right)} \frac{dy}{a}.$$

This being so, we have

$$\mathcal{F}[g](\omega) = \frac{1}{|a|} e^{\frac{i\omega b}{a}} \cdot \frac{1}{2\pi} \int_{-\infty}^{\infty} f(y) e^{-\frac{i\omega y}{a}}\, dy$$

$$= \frac{1}{|a|} e^{\frac{i\omega b}{a}} \mathcal{F}[f]\left(\frac{\omega}{a}\right). \qquad\blacksquare$$

Two special cases of the last formula are:

(a) $b = 0$, i.e., $g(x) = f(ax)$, $a \neq 0$. In this case

$$\mathcal{F}[g](\omega) = \tfrac{1}{|a|}\mathcal{F}[f](\tfrac{\omega}{a})$$

(b) $a = 1$, i.e., $g(x) = f(x + b)$, and then

$$\mathcal{F}[g](\omega) = e^{i\omega b}\mathcal{F}[f](\omega)$$

Because of the many applications of the Fourier transform in physics and electrical engineering, it is usual to render the variable x as time, and the variable ω as frequency. In these terms we can state the last formula as follows: shift in time is equivalent to rotation in frequency.

3.5. For each $f \in G(\mathbb{R})$ and every $c \in \mathbb{R}$

$$\mathcal{F}[e^{icx}f(x)](\omega) = \mathcal{F}[f](\omega - c)$$

Proof:

$$\mathcal{F}[e^{icx}f(x)](\omega) = \frac{1}{2\pi}\int_{-\infty}^{\infty} e^{icx}f(x)e^{-i\omega x}\,dx$$

$$= \frac{1}{2\pi}\int_{-\infty}^{\infty} f(x)e^{-i(\omega - c)x}\,dx$$

$$= \mathcal{F}[f](\omega - c). \qquad\blacksquare$$

This formula may be stated, like the previous formula, in the following terms: rotation in time is equivalent to shift in frequency.

3.6. Modulation Formulae: For each $f \in G(\mathbb{R})$ and every $c \in \mathbb{R}$,

$$\mathcal{F}[f(x)\cos cx](\omega) = \frac{\mathcal{F}[f](\omega - c) + \mathcal{F}[f](\omega + c)}{2}$$

$$\mathcal{F}[f(x)\sin cx](\omega) = \frac{\mathcal{F}[f](\omega - c) - \mathcal{F}[f](\omega + c)}{2i}$$

Proof: We only prove the first formula. The proof of the second is totally analogous. From Properties 3.1 and 3.5

$$\mathcal{F}[f(x)\cos cx](\omega) = \mathcal{F}\left[f(x)\frac{e^{icx} + e^{-icx}}{2}\right](\omega)$$

$$= \frac{1}{2}\mathcal{F}\left[f(x)e^{icx}\right](\omega) + \frac{1}{2}\mathcal{F}\left[f(x)e^{-icx}\right](\omega)$$

$$= \frac{\mathcal{F}[f](\omega - c) + \mathcal{F}[f](\omega + c)}{2}. \blacksquare$$

3.7. Derivative Formula: Let f be a continuous function such that $f, f' \in G(\mathbb{R})$. Then

$$\boxed{\mathcal{F}[f'](\omega) = i\omega \mathcal{F}[f](\omega)}$$

Proof: We first prove that $\lim_{x \to \pm\infty} f(x) = 0$. Since $f' \in G(\mathbb{R})$ and f is continuous, it follows that

$$\lim_{x \to \infty}[f(x) - f(0)] = \lim_{x \to \infty} \int_0^x f'(t)\,dt$$

and the limit of the right-hand side exists (and is finite). Thus $\lim_{x \to \infty} f(x)$ exists. Moreover $f \in G(\mathbb{R})$, and therefore this limiting value must be zero. Thus $\lim_{x \to \infty} f(x) = 0$. Similarly $\lim_{x \to -\infty} f(x) = 0$.

Now

$$\mathcal{F}[f'](\omega) = \frac{1}{2\pi}\int_{-\infty}^{\infty} f'(x)e^{-i\omega x}\,dx = \lim_{L,M \to \infty} \frac{1}{2\pi}\int_{-M}^{L} f'(x)e^{-i\omega x}\,dx.$$

We calculate this last integral by integration by parts:

$$\frac{1}{2\pi}\int_{-M}^{L} f'(x)e^{-i\omega x}\,dx = \frac{1}{2\pi}\left[f(x)e^{-i\omega x}\Big|_{-M}^{L} - \int_{-M}^{L} f(x)(-i\omega)e^{-i\omega x}\,dx\right]$$

$$= \frac{1}{2\pi}\left[f(L)e^{-i\omega L} - f(-M)e^{i\omega M} + i\omega \int_{-M}^{L} f(x)e^{-i\omega x}\,dx\right].$$

Since $\lim_{x \to \pm\infty} f(x) = 0$ we have

$$\lim_{L,M \to \infty}\left[f(L)e^{-i\omega L} - f(-M)e^{i\omega M}\right] = 0.$$

Because $f \in G(\mathbb{R})$ we finally obtain

$$\lim_{L,M \to \infty} \frac{i\omega}{2\pi}\int_{-M}^{L} f(x)e^{-i\omega x}\,dx = i\omega\,\mathcal{F}[f](\omega). \blacksquare$$

In the same manner it may be proven that if $f, f', \ldots, f^{(n-1)}$ are continuous and $f, f', \ldots, f^{(n)} \in G(\mathbb{R})$, then

$$\boxed{\mathcal{F}\!\left[f^{(n)}\right]\!(\omega) = (i\omega)^n \, \mathcal{F}[f](\omega)}$$

Remark: In the remark following Theorem 2.15 we noted a correspondence between the smoothness of a function and the rate at which its Fourier coefficients converge to zero. This same correlation exists, in a general way, between the smoothness of a function and the decay of its Fourier transform at $\pm\infty$. Using the derivative formula we can illustrate one half of this correlation. Under the above assumption $\mathcal{F}\!\left[f^{(n)}\right]\!(\omega) = (i\omega)^n \mathcal{F}[f](\omega)$, since $f^{(n)} \in G(\mathbb{R})$, it follows from Theorem 3.1 that $\mathcal{F}\!\left[f^{(n)}\right]$ is continuous and bounded (in fact more is true). Thus there exists a constant C for which $|\mathcal{F}\!\left[f^{(n)}\right]\!(\omega)| \leq C$ for all $\omega \in \mathbb{R}$. Hence

$$|\mathcal{F}[f](\omega)| \leq \frac{C}{|\omega|^n}$$

for all $\omega \in \mathbb{R}$.

3.8. Let $f \in G(\mathbb{R})$ be such that the integral $\int_{-\infty}^{\infty} |xf(x)|\, dx$ converges. Then the Fourier transform of f is continuously differentiable and satisfies

$$\boxed{\mathcal{F}[xf(x)](\omega) = i\tfrac{d}{d\omega}\mathcal{F}[f](\omega)}$$

Proof: If f is continuous on all of $\mathbb{R}$ then the integral $\int_{-\infty}^{\infty} xf(x)e^{-i\omega x} dx$ converges uniformly on $\mathbb{R}$. Therefore by Leibniz's rule (see Appendix B) we may differentiate with respect to ω under the integration sign,

$$\frac{d}{d\omega}\mathcal{F}[f](\omega) = \frac{d}{d\omega}\frac{1}{2\pi}\int_{-\infty}^{\infty} f(x)e^{-i\omega x} dx = \frac{1}{2\pi}\int_{-\infty}^{\infty} \frac{d}{d\omega}\left[f(x)e^{-i\omega x}\right] dx$$

$$= \frac{1}{2\pi}\int_{-\infty}^{\infty} -ixf(x)e^{-i\omega x} dx = -\frac{i}{2\pi}\int_{-\infty}^{\infty} xf(x)e^{-i\omega x} dx$$

$$= -i\mathcal{F}[xf(x)],$$

to obtain the desired result. If f is only piecewise continuous, then let $\{(a_n, b_n)\}_{n \in I}$ be the family of intervals over which f is continuous (a_n may equal $-\infty$, b_n may equal ∞, and the union of intervals is all of $\mathbb{R}$ minus the points of discontinuity of f). Then

$$\mathcal{F}[f](\omega) = \sum_{n \in I} \frac{1}{2\pi} \int_{a_n}^{b_n} f(x) e^{-i\omega x} dx$$

and the series on the right-hand side converges uniformly on $\mathbb{R}$. We may therefore differentiate $\mathcal{F}[f](\omega)$ in the same way under the sum and the integral signs to obtain our desired result. ∎

In an analogous manner we can prove that if $f \in G(\mathbb{R})$, $n \in \mathbb{N}$, and the integral $\int_{-\infty}^{\infty} |x^n f(x)|\, dx$ converges, then the Fourier transform of f is n times continuously differentiable and satisfies the equality

$$\mathcal{F}[x^n f(x)](\omega) = i^n \frac{d^n}{d\omega^n} \mathcal{F}[f](\omega)$$

Property 3.8 should be compared with the derivative formula (Property 3.7).

Exercises

1. Determine the Fourier transform of each of the following functions:

   (a) $f(x) = e^{-a|x|}$, $a > 0$

   (b) $f(x) = e^{-4x^2 - 4x - 1}$

   (c) $f(x) = 4x e^{-x^2}$

2. Let
$$H(x) = \begin{cases} 0, & x < 0, \\ 1, & x \geq 0. \end{cases}$$

   Find the Fourier transform of

   (a) $f(x) = H(x) e^{-ax}$, $a > 0$

   (b) $f(x) = H(x) e^{-ax} \cos bt$, $a > 0$, $b \neq 0$

   (c) $f(x) = H(x) e^{-ax} \sin bt$, $a > 0$, $b \neq 0$

3. Let $f \in G(\mathbb{R})$. Determine the Fourier transform of

   (a) $f(-x)$

   (b) $f(x - x_0)$, x_0 is a real constant

   (c) $f(x) e^{i\omega_0 x}$, ω_0 is a real constant

   (d) $f(x) \sin \omega_0 x$

   (e) $f(x) \cos \omega_0 x$

   (f) $e^{ix} f(3x)$

   (g) $f(2x)$

4. Assume that f and f' are continuous, $f, f', f'' \in G(\mathbb{R})$, and $xf(x)$ is also absolutely integrable. Assume that f satisfies the differential equation

$$f''(x) + 2x f'(x) + 2f(x) = 0.$$

What differential equation does $\mathcal{F}[f]$ satisfy?

5. Assume that f and f' are continuous, $f, f', f'' \in G(\mathbb{R})$, and $xf(x)$ is also absolutely integrable. Use Fourier transforms to solve the equation

$$f''(x) + xf'(x) + f(x) = 0, \qquad f(0) = 1, \quad f'(0) = 0.$$

6. Let f be a function which is twice differentiable on all of $\mathbb{R}$ such that $f(t)$, $f'(t)$, $f''(t)$, $tf(t)$, and $t^2 f(t)$ are continuous and absolutely integrable over $\mathbb{R}$. We denote the Fourier transform of f by F. Find a real number c such that if

$$f''(t) + (t^2 - 2)f(t) = cf(t),$$

then

$$F''(\omega) + (\omega^2 - 2)F(\omega) = cF(\omega).$$

7. Given $\mathcal{F}[f](\omega) = \frac{1}{1+\omega^2}$, calculate $\mathcal{F}\left[x^2 f''(x) + 2f'''(x)\right](\omega)$.

4. The Inverse Fourier Transform and Plancherel's Identity

The Fourier transform is an example of a mathematical formula taking a function f and producing for us a second function F. (Formulae of this sort are often called *operators*.) In the case of the Fourier transform there exists an additional simple and similar formula which permits us to reverse the process. That is, it takes the Fourier transform and gives us the original function. This operation we call the *inverse Fourier transform*. Formally, this operation is given by the formula

$$f(x) = \int_{-\infty}^{\infty} \mathcal{F}[f](\omega) e^{i\omega x} \, d\omega.$$

Our goal is to understand more precisely why and if and when (under what circumstances) this formula holds. That is, in what exact sense it is the inverse Fourier transform. Before presenting an explanation of these facts, we should point out that different textbooks define the Fourier transform and inverse Fourier transform in different ways. In some books we can find the definition of the Fourier transform to be what we have defined as the inverse Fourier transform (with or without the constant $\frac{1}{2\pi}$ before the integral). In these books the inverse Fourier transform is then given as our Fourier transform. In some books the constant $\frac{1}{\sqrt{2\pi}}$ appears before the integrals in the definition of both the Fourier transform and the inverse Fourier transform. Why all these variants? These differences are purely technical. The only important point is that $e^{-i\omega x}$ appears in one of the formulae, while $e^{i\omega x}$ appears in the other. In addition, the

Chapter 3: The Fourier Transform

product of the constants before the integrals in the two formulae must equal $\frac{1}{2\pi}$.

Theorem 3.3: (**Inverse Fourier Transform**) *If $f \in G(\mathbb{R})$ then for every point $x \in \mathbb{R}$ where its one-sided derivatives exist, we have*

$$\frac{f(x-) + f(x+)}{2} = \lim_{M \to \infty} \int_{-M}^{M} \mathcal{F}[f](\omega) e^{i\omega x} \, d\omega.$$

Proof: By using Fubini's Theorem (see Appendix B)

$$\int_{-M}^{M} \mathcal{F}[f](\omega) e^{i\omega x} \, d\omega = \int_{-M}^{M} \left[\frac{1}{2\pi} \int_{-\infty}^{\infty} f(y) e^{-i\omega y} \, dy \right] e^{i\omega x} \, d\omega$$

$$= \frac{1}{2\pi} \int_{-\infty}^{\infty} f(y) \left[\int_{-M}^{M} e^{-i\omega y} e^{i\omega x} \, d\omega \right] dy$$

$$= \frac{1}{2\pi} \int_{-\infty}^{\infty} f(y) \left[\frac{e^{i\omega(x-y)}}{i(x-y)} \bigg|_{-M}^{M} \right] dy$$

$$= \frac{1}{2\pi} \int_{-\infty}^{\infty} f(y) \frac{2 \sin M(x-y)}{x-y} \, dy$$

$$= \frac{1}{\pi} \int_{x}^{\infty} \frac{f(y) \sin M(x-y)}{x-y} \, dy + \frac{1}{\pi} \int_{-\infty}^{x} \frac{f(y) \sin M(x-y)}{x-y} \, dy.$$

Substituting $t = y - x$ we obtain

$$\int_{-M}^{M} \mathcal{F}[f](\omega) e^{i\omega x} \, d\omega = \frac{1}{\pi} \int_{0}^{\infty} f(x+t) \frac{\sin Mt}{t} \, dt + \frac{1}{\pi} \int_{-\infty}^{0} f(x+t) \frac{\sin Mt}{t} \, dt.$$

We claim that

$$\lim_{M \to \infty} \frac{1}{\pi} \int_{0}^{\infty} f(x+t) \frac{\sin Mt}{t} \, dt = \frac{f(x+)}{2}$$

and

$$\lim_{M \to \infty} \frac{1}{\pi} \int_{-\infty}^{0} f(x+t) \frac{\sin Mt}{t} \, dt = \frac{f(x-)}{2}.$$

We prove only the first result. The second result is proven in the same way (or may be obtained from the first by substitution).

We first separate the integral into the sum of two integrals:

$$\int_{0}^{\infty} f(x+t) \frac{\sin Mt}{t} \, dt = \int_{0}^{\pi} f(x+t) \frac{\sin Mt}{t} \, dt + \int_{\pi}^{\infty} f(x+t) \frac{\sin Mt}{t} \, dt.$$

The choice of π is absolutely arbitrary and has no particular significance. Let us define the function

$$g(t) = \begin{cases} \frac{f(x+t)}{t}, & t > \pi, \\ 0, & t \leq \pi. \end{cases}$$

From the fact that $f \in G(\mathbb{R})$ it also follows that $g \in G(\mathbb{R})$. From the proof of part 3 of Theorem 3.1 we obtain

$$\lim_{M \to \infty} \int_{-\infty}^{\infty} g(t) \sin Mt \, dt = 0.$$

In other words

$$\lim_{M \to \infty} \frac{1}{\pi} \int_{\pi}^{\infty} f(x+t) \frac{\sin Mt}{t} \, dt = 0.$$

We must now deal with the limit

$$\lim_{M \to \infty} \frac{1}{\pi} \int_{0}^{\pi} f(x+t) \frac{\sin Mt}{t} \, dt.$$

The function

$$h(t) = \frac{f(x+t) - f(x+)}{t}$$

is piecewise continuous on $[0, \pi]$. (This property holds at $x = 0$ by the assumption of the theorem.) Again using part 3 of Theorem 3.1 we obtain

$$\lim_{M \to \infty} \frac{1}{\pi} \int_{0}^{\pi} f(x+t) \frac{\sin Mt}{t} \, dt$$

$$= \lim_{M \to \infty} \left[\frac{1}{\pi} \int_{0}^{\pi} \frac{f(x+t) - f(x+)}{t} \sin Mt \, dt + \frac{1}{\pi} f(x+) \int_{0}^{\pi} \frac{\sin Mt}{t} \, dt \right]$$

$$= \frac{1}{\pi} f(x+) \lim_{M \to \infty} \int_{0}^{\pi} \frac{\sin Mt}{t} \, dt.$$

It now remains to verify that the last limit exists and equals $\frac{\pi}{2}$. Substituting $u = Mt$ we obtain

$$\int_{0}^{\pi} \frac{\sin Mt}{t} \, dt = \int_{0}^{M\pi} \frac{\sin u}{u} \, du$$

and thus

$$\lim_{M \to \infty} \int_{0}^{\pi} \frac{\sin Mt}{t} \, dt = \int_{0}^{\infty} \frac{\sin u}{u} \, du$$

assuming that the integral on the right-hand side converges.

Claim: $\int_{0}^{\infty} \frac{\sin u}{u} \, du = \frac{\pi}{2}.$

Proof of the Claim: The function $\frac{\sin u}{u}$ is continuous on the interval $[0, \infty)$, and thus to prove the convergence of the integral it suffices to prove that the limit

$$\lim_{R \to \infty} \int_{a}^{R} \frac{\sin u}{u} \, du$$

exists for any particular $a > 0$. By integration by parts,

$$\int_{a}^{R} \frac{\sin u}{u} \, du = -\frac{\cos u}{u} \Big|_{a}^{R} - \int_{a}^{R} \frac{\cos u}{u^2} \, du.$$

Chapter 3: The Fourier Transform

Obviously

$$\lim_{R \to \infty} \frac{\cos R}{R} = 0$$

and also

$$\left|\frac{\cos u}{u^2}\right| \leq \frac{1}{u^2}.$$

The function $\frac{1}{u^2}$ is absolutely integrable on $[a, \infty)$ for any $a > 0$. Thus our integral does converge.

We have shown that the integral converges. We may therefore obtain its value by considering any limit of the form

$$\lim_{m \to \infty} \int_0^{R_m} \frac{\sin u}{u} du$$

where $\{R_m\}_{m=1}^\infty$ is any sequence going to infinity. Let $R_m = (m+\frac{1}{2})\pi$ for integer m. Thus

$$\int_0^\infty \frac{\sin u}{u} du = \lim_{m \to \infty} \int_0^{(m+\frac{1}{2})\pi} \frac{\sin u}{u} du.$$

Substituting $t = \frac{u}{m+\frac{1}{2}}$ we obtain

$$\int_0^{(m+\frac{1}{2})\pi} \frac{\sin u}{u} du = \int_0^\pi \frac{\sin(m+\frac{1}{2})t}{t} dt.$$

From the proof of Dirichlet's Theorem (Theorem 2.6)

$$\lim_{m \to \infty} \int_0^\pi \frac{\sin(m+\frac{1}{2})t}{t} dt = \lim_{m \to \infty} \int_0^\pi \frac{2\sin\frac{1}{2}t}{t} \cdot \frac{\sin(m+\frac{1}{2})t}{2\sin\frac{1}{2}t} dt$$

$$= \frac{\pi}{2} \lim_{t \to 0+} \frac{2\sin\frac{1}{2}t}{t}$$

$$= \frac{\pi}{2}.$$ ∎

Remark: It may be that $\lim_{M \to \infty} \int_{-M}^M g(x)\, dx$ exists, while $\int_{-\infty}^\infty g(x)\, dx$ does not exist. (Take g to be any odd function which is not integrable.) However, if $\int_{-\infty}^\infty g(x)\, dx$ exists, then $\lim_{M \to \infty} \int_{-M}^M g(x)\, dx$ also exists, and has the same value. We call $\lim_{M \to \infty} \int_{-M}^M g(x)\, dx$, if it exists, the *principal value* of the integral, and denote it by

$$\text{P.V.} \int_{-\infty}^\infty g(x)\, dx.$$

An immediate consequence of Theorem 3.3 is: if $f \in G(\mathbb{R})$, f is continuous, and f' is piecewise continuous, then the following two formulae hold:

$$\mathcal{F}[f](\omega) = \frac{1}{2\pi} \int_{-\infty}^{\infty} f(x) e^{-i\omega x}\, dx,$$

$$f(x) = \text{P.V.} \int_{-\infty}^{\infty} \mathcal{F}[f](\omega) e^{i\omega x}\, d\omega.$$

If we also assume that $\mathcal{F}[f] \in G(\mathbb{R})$, then

$$\mathcal{F}[\mathcal{F}[f](\omega)](x) = \frac{1}{2\pi} \int_{-\infty}^{\infty} \mathcal{F}[f](\omega) e^{-i\omega x}\, d\omega$$

$$= \frac{1}{2\pi} \int_{-\infty}^{\infty} \mathcal{F}[f](\omega) e^{i\omega(-x)}\, d\omega$$

$$= \frac{1}{2\pi} f(-x).$$

In other words, under the above assumptions on f and $\mathcal{F}[f]$ we obtain the formula

$$\mathcal{F}[\mathcal{F}[f]](x) = \frac{1}{2\pi} f(-x).$$

Example 3.4: In Example 3.1 we found the Fourier transform of $f(x) = e^{-|x|}$ to be

$$F(\omega) = \frac{1}{\pi(1+\omega^2)}.$$

Both f and $\mathcal{F}[f]$ are in $G(\mathbb{R})$, continuous, and continuously differentiable. From the inverse Fourier transform formulae we therefore obtain

$$e^{-|x|} = \int_{-\infty}^{\infty} \frac{1}{\pi(1+\omega^2)} e^{i\omega x}\, d\omega$$

$$= \int_{-\infty}^{\infty} \frac{1}{\pi(1+\omega^2)} [\cos \omega x + i \sin \omega x]\, d\omega$$

$$= \int_{-\infty}^{\infty} \frac{\cos \omega x}{\pi(1+\omega^2)}\, d\omega + i \int_{-\infty}^{\infty} \frac{\sin \omega x}{\pi(1+\omega^2)}\, d\omega.$$

The purely imaginary part vanishes since $e^{-|x|}$ is a real function (or because $\frac{\sin \omega x}{\pi(1+\omega^2)}$ is an odd function). This being so, we obtain the formula

$$\frac{2}{\pi} \int_0^{\infty} \frac{\cos \omega x}{1+\omega^2}\, d\omega = e^{-|x|}.$$

Chapter 3: The Fourier Transform

Example 3.5: In Example 3.2 we found the Fourier transform of the function

$$f(x) = \begin{cases} 1, & -b \leq x \leq b, \\ 0, & \text{otherwise,} \end{cases}$$

to be

$$F(\omega) = \frac{\sin \omega b}{\omega \pi}.$$

From the inverse Fourier transform formula and using properties of evenness and oddness, we obtain for $|x| \neq b$,

$$f(x) = \text{P.V.} \int_{-\infty}^{\infty} \frac{\sin \omega b}{\omega \pi} e^{i\omega x} \, d\omega$$

$$= \lim_{M \to \infty} \int_{-M}^{M} \frac{\sin \omega b \cos \omega x}{\omega \pi} \, d\omega + \lim_{M \to \infty} i \int_{-M}^{M} \frac{\sin \omega b \sin \omega x}{\omega \pi} \, d\omega$$

$$= \lim_{M \to \infty} \frac{2}{\pi} \int_0^M \frac{\sin \omega b \cos \omega x}{\omega} \, d\omega = \frac{2}{\pi} \int_0^{\infty} \frac{\sin \omega b \cos \omega x}{\omega} \, d\omega.$$

At the points $x = \pm b$, the function f is not continuous, and thus by Theorem 3.3 the integral converges to the average of the one-sided limits, which is $\frac{1}{2}$. This being so, we obtain the formula

$$\frac{2}{\pi} \int_0^{\infty} \frac{\sin \omega b \cos \omega x}{\omega} \, d\omega = \begin{cases} 1, & |x| < b, \\ \frac{1}{2}, & |x| = b, \\ 0, & |x| > b. \end{cases}$$

By setting $x = 0$ and $b = 1$ (the value of b is in fact irrelevant) we have the previously proven

$$\int_0^{\infty} \frac{\sin \omega}{\omega} \, d\omega = \frac{\pi}{2}.$$

We will now consider two forms of Plancherel's identity. Plancherel's identity is to Fourier transforms as Parseval's identity is to Fourier series. We will not give an exact proof of Plancherel's identity. However, we will try to explain why it is both plausible and true.

Plancherel Identity: *If $f \in G(\mathbb{R})$ and $\int_{-\infty}^{\infty} |f(x)|^2 \, dx < \infty$, then*

$$\int_{-\infty}^{\infty} |\mathcal{F}[f](\omega)|^2 \, d\omega < \infty$$

and in addition,

$$\frac{1}{2\pi} \int_{-\infty}^{\infty} |f(x)|^2 \, dx = \int_{-\infty}^{\infty} |\mathcal{F}[f](\omega)|^2 \, d\omega.$$

This formula is sometimes called the formula for the *conservation of energy*. The left-hand side represents the energy of a signal in the time domain, while the right-hand side represents its energy in the frequency domain. Plancherel's identity is a special case of the following result.

Generalized Plancherel Identity: If $f, g \in G(\mathbb{R})$, $\int_{-\infty}^{\infty} |f(x)|^2 < \infty$, and $\int_{-\infty}^{\infty} |g(x)|^2 < \infty$, then

$$\frac{1}{2\pi} \int_{-\infty}^{\infty} f(x) \overline{g(x)} \, dx = \int_{-\infty}^{\infty} \mathcal{F}[f](\omega) \overline{\mathcal{F}[g](\omega)} \, d\omega.$$

If we set $f = g$ in this formula, then we obtain the previous formula. The validity of the formula is partially justified by the following:

$$\int_{-\infty}^{\infty} \mathcal{F}[f](\omega) \overline{\mathcal{F}[g](\omega)} \, d\omega = \int_{-\infty}^{\infty} \mathcal{F}[f](\omega) \left(\frac{1}{2\pi} \int_{-\infty}^{\infty} g(x) e^{-i\omega x} \, dx \right) d\omega$$

$$= \frac{1}{2\pi} \int_{-\infty}^{\infty} \int_{-\infty}^{\infty} \mathcal{F}[f](\omega) \overline{g(x)} e^{i\omega x} \, dx \, d\omega$$

$$= \frac{1}{2\pi} \int_{-\infty}^{\infty} \left(\int_{-\infty}^{\infty} \mathcal{F}[f](\omega) e^{i\omega x} \, d\omega \right) \overline{g(x)} \, dx$$

$$= \frac{1}{2\pi} \int_{-\infty}^{\infty} f(x) \overline{g(x)} \, dx.$$

A complete proof would be obtained if we justified each of the above steps.

Example 3.6: Let $f(x) = e^{-|x|}$. From Example 3.1, $\mathcal{F}[f](\omega) = \frac{1}{\pi(1+\omega^2)}$. Thus from Plancherel's identity

$$\frac{1}{2\pi} \int_{-\infty}^{\infty} \left(e^{-|x|} \right)^2 dx = \int_{-\infty}^{\infty} \left[\frac{1}{\pi(1+\omega^2)} \right]^2 d\omega.$$

Since $\left(e^{-|x|} \right)^2 = e^{-2|x|}$ and

$$\frac{1}{2\pi} \int_{-\infty}^{\infty} e^{-2|x|} \, dx = \frac{1}{\pi} \int_{0}^{\infty} e^{-2x} \, dx = \frac{1}{2\pi}$$

we have

$$\int_{0}^{\infty} \frac{1}{(1+\omega^2)^2} \, d\omega = \frac{\pi}{4}.$$

Example 3.7: For each constant $b > 0$, let

$$f_b(x) = \begin{cases} 1, & |x| \leq b, \\ 0, & \text{otherwise.} \end{cases}$$

From Example 3.2
$$\mathcal{F}[f_b](\omega) = \frac{\sin \omega b}{\omega \pi}.$$
Choose two constants $a, b > 0$. From the generalized Plancherel identity
$$\frac{1}{2\pi} \int_{-\infty}^{\infty} f_a(x) f_b(x)\, dx = \int_{-\infty}^{\infty} \mathcal{F}[f_a](\omega) \mathcal{F}[f_b](\omega)\, d\omega.$$
The left-hand integral is easily calculated:
$$\frac{1}{2\pi} \int_{-\infty}^{\infty} f_a(x) f_b(x)\, dx = \frac{\min(a,b)}{\pi}.$$
Moreover
$$\int_{-\infty}^{\infty} \mathcal{F}[f_a](\omega) \mathcal{F}[f_b](\omega)\, d\omega = \frac{1}{\pi^2} \int_{-\infty}^{\infty} \frac{\sin a\omega \sin b\omega}{\omega^2}\, d\omega.$$
Thus for $a, b > 0$
$$\int_{-\infty}^{\infty} \frac{\sin a\omega \sin b\omega}{\omega^2}\, d\omega = \pi \min(a,b).$$

Exercises

1. Determine the Fourier transform of each of the following functions:

    (a) $f(x) = \frac{1}{a^2+x^2}$ (b) $f(x) = \frac{\cos ax}{a^2+x^2}$ (c) $f(x) = \frac{\sin bx}{a^2+x^2}$

2. Using results of previous exercises, determine the following integrals:

    (a) $\int_0^\infty \left[\frac{\sin u - u \cos u}{u^2} \right]^2 du$ (b) $\int_0^\infty \frac{1}{(u^2+a^2)^2}\, du$

    (c) $\int_0^\infty \frac{u^2}{(u^2+a^2)^2}\, du$ (d) $\int_0^\infty \frac{x \sin \pi x}{1-x^2}\, dx$

    (e) $\int_0^\infty \frac{x \sin \pi x \cos \pi x}{1-x^2}\, dx$ (f) $\int_0^\infty \left(\frac{x \sin \pi x}{1-x^2} \right)^2 dx$

3. Determine the Fourier transform of
$$f(x) = \begin{cases} 1, & 0 < x \leq a, \\ -1, & -a \leq x \leq 0, \\ 0, & \text{otherwise,} \end{cases}$$
where $a > 0$, and use it to calculate
$$\int_0^\infty \frac{\cos ax - 1}{x} \sin bx\, dx, \qquad b > 0.$$

4. Using the generalized Plancherel identity, calculate the integral
$$\int_0^\infty \frac{dt}{(a^2+t^2)(b^2+t^2)}, \qquad a, b > 0.$$

5. Let $f : \mathbb{R} \to \mathbb{C}$ be continuous and absolutely integrable on $\mathbb{R}$. Let F be the Fourier transform of f. It is known that

$$F(\omega) = \begin{cases} 1 - \omega^2, & |\omega| \leq 1, \\ 0, & |\omega| > 1. \end{cases}$$

Find f.

6. The same exercise as Exercise 5, except that

$$F(\omega) = \begin{cases} 1 - \omega, & |\omega| \leq 1, \\ 0, & |\omega| > 1. \end{cases}$$

5. Convolution

Let f and g be two given functions. There are various mathematical operations which take two functions and form a new function. Consider, for example, $f + g$, $f \cdot g$, and $f(g(\cdot))$ (composition). In this section we define a new operation which is of central importance in the subject of Fourier transforms, and almost any integral transform. This operation is called *convolution* and we denote it by $f * g$. Let f and g be functions whose domain of definition is all of $\mathbb{R}$. For each $x \in \mathbb{R}$, we define

$$(f * g)(x) = \int_{-\infty}^{\infty} f(x - y) g(y) \, dy$$

if the integral exists. Under the assumption that the integral (convolution) exists, then by substitution it is readily verified that

$$(f * g)(x) = \int_{-\infty}^{\infty} f(y) g(x - y) \, dy.$$

In other words

$$f * g = g * f.$$

The operation of convolution is commutative. There are a number of further basic properties of convolution which we will remark upon as needed. One such is the following.

Proposition 3.4: *If $f, g \in G(\mathbb{R})$, then the convolution $f * g$ exists and is absolutely integrable.*

Proof: It follows from Fubini's Theorem (Theorem B.4) that

$$\int_{-\infty}^{\infty} |(f*g)(x)|\, dx = \int_{-\infty}^{\infty} \left| \int_{-\infty}^{\infty} f(x-y)g(y)\, dy \right| dx$$

$$\leq \int_{-\infty}^{\infty} \int_{-\infty}^{\infty} |f(x-y)| \cdot |g(y)|\, dy\, dx$$

$$= \int_{-\infty}^{\infty} \left[\int_{-\infty}^{\infty} |f(x-y)|\, dx \right] \cdot |g(y)|\, dy.$$

We assumed that f is absolutely integrable on $\mathbb{R}$. Thus by substitution we obtain

$$\int_{-\infty}^{\infty} |f(x-y)|\, dx = \int_{-\infty}^{\infty} |f(x)|\, dx < \infty$$

and the value of the integral is independent of y. Thus

$$\int_{-\infty}^{\infty} |(f*g)(x)|\, dx \leq \int_{-\infty}^{\infty} \left[\int_{-\infty}^{\infty} |f(x)|\, dx \right] |g(y)|\, dy$$

$$= \left(\int_{-\infty}^{\infty} |f(x)|\, dx \right) \left(\int_{-\infty}^{\infty} |g(y)|\, dy \right) < \infty. \qquad \blacksquare$$

The importance of convolution, in the context of Fourier transform, is contained in this next result.

Theorem 3.5: (Convolution Theorem) *For each* $f, g \in G(\mathbb{R})$

$$\mathcal{F}[f*g](\omega) = 2\pi \mathcal{F}[f](\omega) \cdot \mathcal{F}[g](\omega).$$

Proof: Since $f, g \in G(\mathbb{R})$ it follows from Proposition 3.4 that $f*g$ is absolutely integrable. It is not, however, necessarily true that $f*g$ is piecewise continuous. Nevertheless, applying Fubini's Theorem (Theorem B.4), one can perform the following operations:

$$\mathcal{F}[f*g](\omega) = \frac{1}{2\pi} \int_{-\infty}^{\infty} (f*g)(x) e^{-i\omega x}\, dx$$

$$= \frac{1}{2\pi} \int_{-\infty}^{\infty} \left(\int_{-\infty}^{\infty} f(x-y)g(y)\, dy \right) e^{-i\omega x}\, dx$$

$$= \frac{1}{2\pi} \int_{-\infty}^{\infty} \int_{-\infty}^{\infty} f(x-y) e^{-i\omega(x-y)} g(y) e^{-i\omega y}\, dx\, dy$$

$$= \int_{-\infty}^{\infty} \left(\frac{1}{2\pi} \int_{-\infty}^{\infty} f(x-y) e^{-i\omega(x-y)}\, dx \right) g(y) e^{-i\omega y}\, dy.$$

It is easily checked (by substitution) that

$$\frac{1}{2\pi}\int_{-\infty}^{\infty} f(x-y)e^{-i\omega(x-y)}\,dx = \mathcal{F}[f](\omega).$$

Since this integral is independent of y, we obtain

$$\mathcal{F}[f*g](\omega) = \mathcal{F}[f](\omega)\int_{-\infty}^{\infty} g(y)e^{-i\omega y}\,dy = \mathcal{F}[f](\omega)\cdot 2\pi\cdot\mathcal{F}[g](\omega).$$
∎

Theorem 3.5 points out the importance of convolution for the Fourier transform. The Fourier transform of the convolution of two functions is equivalent (up to a constant) to the multiplication of their Fourier transforms. In general we use this result in the "reverse" direction. That is, we are given $\mathcal{F}[f](\omega)\cdot\mathcal{F}[g](\omega)$, where the functions f and g are known, and the object is to find a function h whose Fourier transform is $\mathcal{F}[h](\omega) = \mathcal{F}[f](\omega)\cdot\mathcal{F}[g](\omega)$. In other words, we wish to find the inverse Fourier transform of $\mathcal{F}[f](\omega)\cdot\mathcal{F}[g](\omega)$. From Theorem 3.5 we know that it is $h(x) = \frac{1}{2\pi}(f*g)(x)$.

Exercises

1. By using the Fourier transform, solve the following integral equations:

   (a) $\int_{-\infty}^{\infty} f(t)f(x-t)\,dt = e^{-\frac{x^2}{2}}$

   (b) $\int_{-\infty}^{\infty} \frac{f(t)}{(x-t)^2+a^2}\,dt = \frac{1}{x^2+b^2}$, $\quad 0 < a < b$

   (c) $\int_{-\infty}^{\infty} f(t)f(x-t)\,dt = \frac{1}{x^2+1}$

   (d) $\int_{-\infty}^{\infty} e^{-|x-t|}f(t)\,dt = e^{-|x|} + |x|e^{-|x|}$.

2. For each $a > 0$ we define the functions

$$f_a(x) = \begin{cases} 1, & |x| < a, \\ 0, & |x| \geq a, \end{cases} \qquad g_a(x) = \begin{cases} 1 - \frac{|x|}{a}, & |x| < a, \\ 0, & |x| \geq a. \end{cases}$$

   Calculate $f_a * f_a$ and $g_a * g_a$.

3. Let $f : \mathbb{R} \to \mathbb{C}$ be a continuous, absolutely integrable function. Let F denote the Fourier transform of f. Given that

$$F(\omega) + \int_{-\infty}^{\infty} F(\omega - s)e^{-|s|}\,ds = \begin{cases} \omega^2, & 0 \leq \omega \leq 1, \\ 0, & \text{otherwise,} \end{cases}$$

   find f.

4. Let $f : \mathbb{R} \to \mathbb{C}$ be a continuous, absolutely integrable function. Let F denote the Fourier transform of f. Show that if $F(\omega) = 0$ for all $\omega > |\omega_0|$ then for all $a > |\omega_0|$
$$f(x) * \frac{\sin ax}{\pi x} = f(x).$$

5. Let F denote the Fourier transform of the function
$$f(x) = \begin{cases} 1, & 0 \leq x < 1, \\ 0, & \text{otherwise.} \end{cases}$$

Find the function g such that the Fourier transform G of g satisfies $G(\omega) = [F(\omega)]^2$.

6. Applications of the Residue Theorem

In this section we show how we can use the Residue Theorem to calculate Fourier transforms.

We first illustrate the general method by example. We will then show how to refine the method so that it applies to a larger class of functions than might have been supposed at first glance.

Example 3.8: Let $f(x) = \frac{1}{x^4+1}$. Obviously $f \in G(\mathbb{R})$. We wish to determine its Fourier transform
$$F(\omega) = \frac{1}{2\pi} \int_{-\infty}^{\infty} \frac{1}{x^4+1} e^{-i\omega x} dx.$$

We will calculate $F(\omega)$ separately for $\omega \geq 0$ and for $\omega \leq 0$.

$\boldsymbol{\omega \geq 0}$: Consider the simple closed curve $\gamma_R = I_R \cup C_R$, where
$$I_R = \{z \mid z = x, \quad -R \leq x \leq R\},$$
$$C_R = \{z \mid z = Re^{i\theta}, \quad \pi \leq \theta \leq 2\pi\}.$$

γ_R is the lower half-circle of radius R centred at 0. We consider γ_R as having positive orientation. Thus
$$\oint_{\gamma_R} = -\int_{I_R} + \int_{C_R}.$$

The function $\frac{e^{-i\omega z}}{z^4+1}$ has four simple poles at $\frac{\pm 1 \pm i}{\sqrt{2}}$. From the Residue Theorem (Theorem A.3), for each $R > 1$,
$$\frac{1}{2\pi i} \oint_{\gamma_R} \frac{e^{-i\omega z}}{z^4+1} dz = \text{Res}\left\{\frac{e^{-i\omega z}}{z^4+1}; \frac{1-i}{\sqrt{2}}\right\} + \text{Res}\left\{\frac{e^{-i\omega z}}{z^4+1}; \frac{-1-i}{\sqrt{2}}\right\}.$$

A calculation shows the right-hand side to equal

$$\frac{ie^{-\frac{\omega}{\sqrt{2}}}}{2\sqrt{2}}\left(\cos\frac{\omega}{\sqrt{2}} + \sin\frac{\omega}{\sqrt{2}}\right).$$

Now,

$$\frac{1}{2\pi i}\oint_{\gamma_R}\frac{e^{-i\omega z}}{z^4+1}dz = -\frac{1}{2\pi i}\int_{-R}^{R}\frac{e^{-i\omega x}}{x^4+1}dx + \frac{1}{2\pi i}\int_{C_R}\frac{e^{-i\omega z}}{z^4+1}dz.$$

C_R is of length πR. On C_R, for $z = x + iy$,

$$|e^{-i\omega z}| = e^{\omega y} \leq 1$$

since $\omega \geq 0$ and $y \leq 0$. Furthermore,

$$\frac{1}{|z^4+1|} \leq \frac{1}{|z|^4-1} = \frac{1}{R^4-1}.$$

Thus

$$\left|\frac{1}{2\pi i}\int_{C_R}\frac{e^{-i\omega z}}{z^4+1}dz\right| \leq \frac{1}{2\pi}\int_{C_R}\frac{|e^{-i\omega z}|}{|z^4+1|}|dz| \leq \frac{1}{2\pi}\cdot\frac{\pi R}{R^4-1} = \frac{R}{2(R^4-1)}$$

which tends to zero as R tends to infinity. Therefore, for $\omega \geq 0$,

$$F(\omega) = \lim_{R\to\infty}\frac{1}{2\pi}\int_{-R}^{R}\frac{e^{-i\omega x}}{x^4+1}dx$$

$$= -i\left[\frac{ie^{-\frac{\omega}{\sqrt{2}}}}{2\sqrt{2}}\left(\cos\frac{\omega}{\sqrt{2}} + \sin\frac{\omega}{\sqrt{2}}\right)\right]$$

$$= \frac{e^{-\frac{\omega}{\sqrt{2}}}}{2\sqrt{2}}\left(\cos\frac{\omega}{\sqrt{2}} + \sin\frac{\omega}{\sqrt{2}}\right).$$

$\omega \leq 0$: Here we do essentially the same calculations. The difference is that in order to bound $|e^{-i\omega z}|$ we use the upper half-circle of radius R centred at 0. That is, $\gamma'_R = I'_R \cup C'_R$, where $I'_R = I_R$ and

$$C'_R = \left\{z \mid z = Re^{i\theta},\ 0 \leq \theta \leq \pi\right\}.$$

Here

$$\oint_{\gamma'_R} = \int_{I'_R} + \int_{C'_R},$$

and the two poles of $\frac{e^{-i\omega z}}{z^4+1}$ which lie in the interior of γ'_R, for every $R > 1$, are at $\frac{\pm 1 + i}{\sqrt{2}}$. The same calculations as above lead us to

$$F(\omega) = \frac{e^{\frac{\omega}{\sqrt{2}}}}{2\sqrt{2}}\left(\cos\frac{\omega}{\sqrt{2}} - \sin\frac{\omega}{\sqrt{2}}\right)$$

for $\omega \leq 0$. (Since f is real and even it follows from Property 3.3 that F is real and even. Thus we could have, in this case, obtained the above formula without calculation.)

Let us now try to understand how we might use the method in Example 3.8 to calculate other Fourier transforms. We must require two properties from our function f. Firstly f should have an extension to the complex plane which is analytic except at a finite number of singular points. Secondly we must be able to verify that for $\omega \geq 0$

$$\lim_{R \to \infty} \int_{C_R} f(z) e^{-i\omega z} dz = 0,$$

and for $\omega \leq 0$

$$\lim_{R \to \infty} \int_{C'_R} f(z) e^{-i\omega z} dz = 0.$$

If these two properties hold then we can calculate

$$\text{P.V.} \int_{-\infty}^{\infty} f(x) e^{-i\omega x} dx$$

for all $\omega \in \mathbb{R}$. If f is absolutely integrable then this is exactly $F(\omega)$.

Both C_R and C'_R are of length πR. For $\omega \geq 0$ we have $|e^{-i\omega z}| \leq 1$ on C_R, and for $\omega \leq 0$ we have $|e^{-i\omega z}| \leq 1$ on C'_R. Thus it follows that our second requirement is valid if

$$\lim_{R \to \infty} \left(R \cdot \max_{z \in C_R \cup C'_R} |f(z)| \right) = 0.$$

This we verified in our example.

However, a weaker condition actually suffices.

Proposition 3.6: *Let C_R and C'_R be as above. If*

$$\lim_{R \to \infty} \max_{z \in C_R \cup C'_R} |f(z)| = 0$$

then for $\omega > 0$

$$\lim_{R \to \infty} \int_{C_R} f(z) e^{-i\omega z} dz = 0$$

and for $\omega < 0$

$$\lim_{R \to \infty} \int_{C'_R} f(z) e^{-i\omega z} dz = 0.$$

The proof of Proposition 3.6 relies on Jordan's Lemma.

Lemma 3.7: (Jordan's Lemma) *For $A > 0$*

$$\int_0^\pi e^{-A\sin\theta}d\theta < \frac{\pi}{A}.$$

Proof: The function $\sin\theta$ is an even function about the point $\frac{\pi}{2}$. Thus it suffices to prove that

$$\int_0^{\frac{\pi}{2}} e^{-A\sin\theta}d\theta < \frac{\pi}{2A}.$$

For each $0 \leq \theta \leq \frac{\pi}{2}$ we have the inequality

$$\frac{2\theta}{\pi} \leq \sin\theta.$$

One justification for this inequality is that the graph of $\frac{2\theta}{\pi}$ on the interval $[0, \frac{\pi}{2}]$ is a straight line connecting the points $(0,0)$ and $(\frac{\pi}{2}, 1)$. The graph of $\sin\theta$ on the same interval has the same endpoints but is concave. Therefore it lies above the graph of $\frac{2\theta}{\pi}$. From this inequality we have

$$\int_0^{\frac{\pi}{2}} e^{-A\sin\theta}d\theta \leq \int_0^{\frac{\pi}{2}} e^{-\frac{2A\theta}{\pi}}d\theta = -\frac{\pi}{2A}e^{-\frac{2A\theta}{\pi}}\Big|_0^{\frac{\pi}{2}} = \frac{\pi}{2A}(1-e^{-A}) < \frac{\pi}{2A}. \blacksquare$$

Proof of Proposition 3.6: We assume $\omega > 0$. The case $\omega < 0$ is proven analogously. For $z \in C_R$, $z = Re^{i\theta}$, $\pi \leq \theta \leq 2\pi$, and

$$|e^{-i\omega z}| = |e^{-i\omega Re^{i\theta}}| = e^{\omega R\sin\theta}.$$

Thus

$$\left|\int_{C_R} f(z)e^{-i\omega z}dz\right| \leq \int_{C_R} |f(z)| \cdot |e^{-i\omega z}| \cdot |dz|$$

$$\leq \max_{z \in C_R}|f(z)| \int_{C_R} |e^{-i\omega z}| \cdot |dz|$$

$$\leq \max_{z \in C_R}|f(z)| \int_\pi^{2\pi} e^{\omega R\sin\theta} R\,d\theta.$$

In the last step we used the fact that $z = Re^{i\theta}$ to obtain $dz = iRe^{i\theta}d\theta$ and $|dz| = Rd\theta$. The substitution $\varphi = \theta - \pi$ leads to

$$\int_\pi^{2\pi} e^{\omega R\sin\theta} \cdot R\,d\theta = R\int_0^\pi e^{-\omega R\sin\varphi}d\varphi.$$

Chapter 3: The Fourier Transform

Hence, since $\omega R > 0$, using Jordan's Lemma

$$\left| \int_{C_R} f(z)e^{-i\omega z} dz \right| \leq \max_{z \in C_R} |f(z)| R \int_0^\pi e^{-wR\sin\varphi} d\varphi$$

$$< \max_{z \in C_R} |f(z)| R \cdot \frac{\pi}{\omega R}$$

$$= \frac{\pi}{\omega} \max_{z \in C_R} |f(z)|.$$

Thus if
$$\lim_{R \to \infty} \max_{z \in C_R} |f(z)| = 0,$$
then
$$\lim_{R \to \infty} \int_{C_R} f(z)e^{-i\omega z} dz = 0$$

for each $\omega > 0$. ∎

We can now summarize this technique for calculating Fourier transforms as follows.

Theorem 3.8: *Assume f is analytic on all of $\mathbb{C}$ except at the points $z_1, \ldots, z_m$, in the upper half-plane, and the points $z_{m+1}, \ldots, z_n$, in the lower half-plane. Assume, in addition, that f is absolutely integrable on $\mathbb{R}$, and*

(3.1) $$\lim_{R \to \infty} \max_{z \in C_R \cup C'_R} |f(z)| = 0.$$

Then F, the Fourier transform of f, is given by

$$F(\omega) = -i \sum_{j=m+1}^n \operatorname{Res}\{f(z)e^{-i\omega z}; z_j\}$$

for $\omega \geq 0$, and

$$F(\omega) = i \sum_{j=1}^m \operatorname{Res}\{f(z)e^{-i\omega z}; z_j\}$$

for $\omega \leq 0$.

Example 3.9: Let $f(x) = \frac{1}{x^2+1}$. The analytic extension of f to all of $\mathbb{C}$ is $f(z) = \frac{1}{z^2+1}$. f has simple poles at $\pm i$, and it is easily verified that f satisfies the other conditions of Theorem 3.8. Thus the Fourier transform F of f is given by

$$F(\omega) = -i \operatorname{Res}\{\frac{e^{-i\omega z}}{z^2+1}; -i\} = -i \cdot \frac{1}{-2i} e^{-i\omega(-i)} = \frac{e^{-\omega}}{2}$$

for $\omega \geq 0$, and
$$F(\omega) = i \operatorname{Res}\{\frac{e^{-i\omega z}}{z^2+1}; i\} = i \cdot \frac{1}{2i} e^{-i\omega i} = \frac{e^\omega}{2}$$
for $\omega \leq 0$. That is,
$$F(\omega) = \frac{1}{2} e^{-|\omega|}.$$

If f has an extension to all of $\mathbb{C}$, which is analytic at all but a finite number of points, then this extension is unique. That is, there is only one such extension. It is also generally easy to guess what this extension must be. However, there is no reason to suppose that this extension satisfies (3.1). For example, if $f(x) = \frac{\sin x}{x^2+1}$, then $f(z) = \frac{\sin z}{z^2+1}$ which does not satisfy (3.1). (From Example 3.9 and the modulation formula (Property 3.6) we can nevertheless easily calculate its Fourier transform.)

f is a *rational function* if f has the form
$$f(x) = \frac{p(x)}{q(x)}$$
where both p and q are polynomials. Its extension to all of $\mathbb{C}$, which is analytic at all but a finite number of points, is given simply by
$$f(z) = \frac{p(z)}{q(z)}.$$
Assume p and q have no common zeros. Then f is absolutely integrable on $\mathbb{R}$ if and only if q has no zero on $\mathbb{R}$, and $2 + \deg(p) \leq \deg(q)$ (where $\deg(p)$ stands for the degree of the polynomial p). We can identify the class of rational functions for which (3.1) holds.

Proposition 3.9: *Assume*
$$f(z) = \frac{p(z)}{q(z)}$$
where p and q are polynomials. Then
$$\lim_{R \to \infty} \max_{z \in C_R \cup C'_R} |f(z)| = 0$$
if and only if
$$\deg(p) < \deg(q).$$

Proof: Let $\deg(q) - \deg(p) = k > 0$. Then there exist $M > 0$ and $c > 0$ such that
$$\left|\frac{p(z)}{q(z)}\right| \leq \frac{c}{|z|^k}$$
for all $|z| \geq M$. This proves the proposition. ∎

Exercises

1. Calculate the Fourier transform of each of the following rational functions:

   (a) $\frac{x^2+1}{x^4+1}$ (b) $\frac{x^3}{x^8+1}$ (c) $\frac{1}{x^6+1}$

   (d) $\frac{1}{(x-a)^2+b^2}$ (e) $\frac{x}{(x^2-2x+2)^2}$ (f) $\frac{x}{(x^2+a^2)(x^2+b^2)}$

2. By using your answers to Exercise 1, compute the following integrals:

   (a) $\int_0^\infty \frac{(x^2+1)\cos 6x}{x^4+1}\,dx$ (b) $\int_{-\infty}^\infty \frac{x \sin mx}{(x^2+a^2)(x^2+b^2)}\,dx$ (c) $\int_0^\infty \frac{\cos mx}{x^6+1}\,dx$

3. Compute the Fourier transform of $f(x) = \frac{1}{e^x+e^{-x}}$. (Hint: Use the simple closed curve $\gamma_R = I_R \cup J_R \cup I'_R \cup J'_R$, where

$$I_R = \{z \mid z = x, \quad -R \leq x \leq R\},$$

$$J_R = \{z \mid z = R + iy, \quad 0 \leq y \leq \pi\},$$

$$I'_R = \{z \mid z = x + i\pi, \quad -R \leq x \leq R\},$$

$$J'_R = \{z \mid z = -R + iy, \quad 0 \leq y \leq \pi\},$$

with positive orientation, and let $R \to \infty$.)

7. Applications to Partial Differential Equations

In this section we give examples of some applications of Fourier transforms to the solutions of certain classical problems in partial differential equations. We assume that our reader has some interest in or prior knowledge of the area. We will not go into the details of the analysis, so those readers who do not fulfil either of the above two requirements should feel free to either read or skip this section. We recommend the former.

The Heat Equation

The one-dimensional homogeneous heat equation on the real line is given by the equation

$$u_t - ku_{xx} = 0, \quad -\infty < x < \infty, \quad 0 < t < \infty.$$

Here u is a function of x and t. The variable x represents a point on the real line and t is time. The real line is a model for a rod of infinite length (in both directions) made of some material of uniform density. We assume that the rod is insulated. This being so, heat does not escape from or enter the rod. The value $u(x,t)$ is the temperature of the rod at the point (or cross-section) x at time t. k is a positive constant which has a physical meaning and depends upon the material of the rod and its density. This equation is a simple mathematical model for the distribution of heat, along a rod, over time. In order to determine a particular solution it is necessary to know the temperature at every point of the rod at some fixed time. This being so, we assume that we are given an initial condition of the form

$$u(x,0) = f(x), \qquad -\infty < x < \infty,$$

for some $f \in G(\mathbb{R})$. We adopt the not unreasonable assumption that the distribution of the temperature of the rod is piecewise continuous, and that the amount of heat in the rod is finite. We also assume that these same conditions hold for u, u_x, u_t, and u_{xx} (as functions of x for each fixed t).

The general idea behind our solution to this problem is that we do not attack u directly. Rather we take the Fourier transform of u (with respect to the variable x, where we consider t as a parameter) and call it U. We then translate our original problem into this new setting. It turns out to be a simpler problem. One reason is that when we apply the Fourier transform to u_{xx} the partial derivatives will disappear (see the derivative Property 3.7). In our case, applying the Fourier transform to the heat equation leads us to an ordinary differential equation of first order for U. The price we pay is twofold. Firstly the coefficients of this ordinary differential equation are not constants. Secondly we solve for U, but not for u. We will need to use the inverse Fourier transform to obtain u.

So let us define

$$U(\omega, t) = \frac{1}{2\pi} \int_{-\infty}^{\infty} u(x,t) e^{-i\omega x}\, dx.$$

In other words $U(\omega, t)$ is the Fourier transform of $u(x,t)$ with respect to the variable x. Differentiate U with respect to t:

$$U_t(\omega, t) = \frac{1}{2\pi} \int_{-\infty}^{\infty} u_t(x,t) e^{-i\omega x}\, dx.$$

Since $u_t = k u_{xx}$ (by the heat equation), we have

$$U_t(\omega, t) = \frac{k}{2\pi} \int_{-\infty}^{\infty} u_{xx}(x,t) e^{-i\omega x}\, dx.$$

From the derivative formula for Fourier transforms (formula 3.7) we have that every twice differentiable $g \in G(\mathbb{R})$ (with some additional conditions) satisfies

$$\mathcal{F}[g''](\omega) = -\omega^2 \mathcal{F}[g](\omega).$$

Chapter 3: The Fourier Transform

Thus (paying little attention to the details) we obtain

$$U_t(\omega, t) = k\mathcal{F}[u_{xx}](\omega, t) = -k\omega^2 U(\omega, t).$$

We have therefore obtained the simple equation

$$U_t + k\omega^2 U = 0.$$

Note that the only derivatives appearing are with respect to t. This being so, this equation can be treated like an ordinary differential equation. It is easy to check that the general solution to this equation is

$$U(\omega, t) = A(\omega) e^{-k\omega^2 t}$$

where $A(\omega)$ is a totally arbitrary function of ω. To determine $A(\omega)$ we use the initial conditions on u:

$$A(\omega) = U(\omega, 0) = \frac{1}{2\pi} \int_{-\infty}^{\infty} u(x, 0) e^{-i\omega x}\, dx = \frac{1}{2\pi} \int_{-\infty}^{\infty} f(x) e^{-i\omega x}\, dx.$$

Thus

$$A(\omega) = F(\omega)$$

where F is the Fourier transform of f (which was given). We have therefore proven that

$$U(\omega, t) = F(\omega) e^{-k\omega^2 t}.$$

It remains to take the inverse Fourier transform of U to obtain u, the solution to our heat equation. We know that F is the Fourier transform of f. If we can calculate p, whose Fourier transform (with respect to x) is $e^{-k\omega^2 t}$, then we can use Theorem 3.5 to obtain u.

So let us determine p. In Example 3.3 we proved that the Fourier transform of $g(x) = e^{-x^2}$ is $\mathcal{F}[g](\omega) = \frac{1}{2\sqrt{\pi}} e^{-\frac{\omega^2}{4}}$. In addition, we also have the shift formula 3.4 which says that if $h(x) = cg(ax)$, $a \neq 0$, then

$$\mathcal{F}[h](\omega) = \frac{c}{|a|} \mathcal{F}[g]\left(\frac{\omega}{a}\right).$$

For $g(x) = e^{-x^2}$ this translates into

$$\mathcal{F}\left[ce^{-a^2 x^2}\right](\omega) = \frac{c}{2\sqrt{\pi}|a|} e^{-\frac{\omega^2}{4a^2}}.$$

We are interested in choosing our constants a and c so that

$$\frac{c}{2\sqrt{\pi}|a|} e^{-\frac{\omega^2}{4a^2}} = e^{-k\omega^2 t}.$$

From this equality it easily follows that $a = \frac{1}{2\sqrt{kt}}$ and $c = \sqrt{\frac{\pi}{kt}}$. Thus if we take

$$p(x,t) = \sqrt{\frac{\pi}{kt}} e^{-\frac{x^2}{4kt}}$$

then

$$\mathcal{F}[p](\omega, t) = \frac{1}{2\pi} \int_{-\infty}^{\infty} p(x,t) e^{-i\omega x}\, dx = e^{-k\omega^2 t}.$$

This being so, from the Convolution Theorem (Theorem 3.5) we have

$$u(x,t) = \frac{1}{2\pi}(f * p)(x,t)\, dx$$

$$= \frac{1}{2\pi} \int_{-\infty}^{\infty} f(y) p(x-y, t)\, dy$$

$$= \frac{1}{2\sqrt{\pi kt}} \int_{-\infty}^{\infty} f(y) e^{-\frac{(x-y)^2}{4kt}}\, dy.$$

This is the solution to this heat equation. The function $e^{-\frac{x^2}{4kt}}$ is sometimes called the *heat equation kernel* or the *Gauss kernel*. Note that we do not in this solution of the heat equation really need to calculate F, the Fourier transform of f. The Fourier transform is but a tool via which we arrived at a solution.

Laplace's Equation on the Half-Plane

We will now consider Laplace's equation on the upper half-plane. It is given by

$$u_{xx} + u_{yy} = 0, \qquad -\infty < x < \infty, \quad 0 < y.$$

Our aim is to find a solution $u(x,y)$ which is continuous for all x and $y \geq 0$, satisfying the boundary condition

$$u(x,0) = f(x), \qquad -\infty < x < \infty,$$

where f is a given function. This condition is not, in itself, sufficient to uniquely determine a solution. During the discussion we will impose an additional condition of boundedness on our solution.

As with the heat equation, we define the Fourier transforms

$$U(\omega, y) = \frac{1}{2\pi} \int_{-\infty}^{\infty} u(x,y) e^{-i\omega x}\, dx$$

and

$$F(\omega) = \mathcal{F}[f](\omega) = \frac{1}{2\pi} \int_{-\infty}^{\infty} f(x) e^{-i\omega x}\, dx = U(\omega, 0).$$

Chapter 3: The Fourier Transform

We now differentiate $U(\omega, y)$ twice with respect to y:

$$U_{yy}(\omega, y) = \frac{1}{2\pi} \int_{-\infty}^{\infty} u_{yy}(x, y) e^{-i\omega x} \, dx.$$

From Laplace's equation $u_{yy} = -u_{xx}$, and thus

$$U_{yy}(\omega, y) = -\frac{1}{2\pi} \int_{-\infty}^{\infty} u_{xx}(x, y) e^{-i\omega x} \, dx.$$

Using the derivative formula 3.7 twice, we obtain

$$U_{yy}(\omega, y) = \omega^2 U(\omega, y).$$

So by use of the Fourier transform we have replaced Laplace's equation $u_{xx} + u_{yy} = 0$ by a new equation which only contains derivatives with respect to y. It is

$$U_{yy} - \omega^2 U = 0.$$

The boundary condition $u(x, 0) = f(x)$ is translated into the boundary condition $U(\omega, 0) = F(\omega)$.

It is readily checked that the general solution to the above equation is

$$U(\omega, y) = A(\omega) e^{|\omega| y} + B(\omega) e^{-|\omega| y}.$$

From the boundedness condition it follows that $A(\omega) = 0$. Thus

$$U(\omega, y) = B(\omega) e^{-|\omega| y}.$$

If we set $y = 0$ we see that

$$F(\omega) = U(\omega, 0) = B(\omega).$$

Thus

$$U(\omega, y) = F(\omega) e^{-|\omega| y}.$$

As in our solution to the heat equation, we have also obtained the Fourier transform of our solution u in the form of a product of two functions of ω. The first function F is simply the Fourier transform of the given function f. The second function is $e^{-|\omega| y}$. If we can find a function p whose Fourier transform with respect to x is $e^{-|\omega| y}$, then from the Convolution Theorem (Theorem 3.5) we will obtain

$$u(x, y) = \frac{1}{2\pi} \int_{-\infty}^{\infty} f(t) \, p(x - t, y) \, dt.$$

Recall that in Example 3.9 we found the Fourier transform of $g(x) = \frac{1}{x^2+1}$ and it is $\mathcal{F}[g](\omega) = \frac{1}{2}e^{-|\omega|}$ (see also Examples 3.1 and 3.4). From the shift formula 3.4 we have that if $h(x) = c \cdot g(ax)$, where $a \neq 0$, then

$$\mathcal{F}[h](\omega) = \frac{c}{|a|}g(\tfrac{\omega}{a}) = \frac{c}{|a|}e^{-|\frac{\omega}{a}|}.$$

We are interested in determining a and c so that

$$\frac{c}{2|a|}e^{-|\frac{\omega}{a}|} = e^{-|\omega|y}.$$

We must therefore choose $a = \frac{1}{y}$ and $c = \frac{2}{y}$ (recall that $y > 0$). Thus

$$p(x,y) = c \cdot g(ax) = c\frac{1}{a^2x^2+1} = \frac{2}{y} \cdot \frac{1}{\frac{x^2}{y^2}+1} = \frac{2y}{x^2+y^2}.$$

We can now write our solution. It is

$$u(x,y) = \frac{y}{\pi}\int_{-\infty}^{\infty}\frac{f(t)}{(x-t)^2+y^2}\,dt.$$

It is possible to directly verify (substitute u in the equation) that this is in fact a solution to Laplace's equation. (But this is not a trivial task.) This particular solution to Laplace's equation on the upper half-plane $y > 0$ is called *Poisson's formula*.

8. Applications to Signal Processing

A signal (function) f is said to be *time-limited* if there exists a constant M such that $f(x) = 0$ for all $|x| \geq M$. A signal (function) $f \in G(\mathbb{R})$ is said to be *band-limited* if there exists a constant L such that $\mathcal{F}[f](\omega) = 0$ for all $|\omega| \geq L$.

In this section we discuss three problems associated with band-limited functions.

Low-Pass Filters

Let $f \in G(\mathbb{R})$, and F denote the Fourier transform of f. Choose a positive constant $L > 0$, and define the function

$$F_L(\omega) = \begin{cases} F(\omega), & |\omega| \leq L, \\ 0, & |\omega| > L. \end{cases}$$

Chapter 3: The Fourier Transform

We say that the function F_L is obtained by passing the function F through a low-pass filter. (Not everyone agrees with this definition, but let us not quibble about it here.) The question we wish to consider is the following. What is the function f_L such that $\mathcal{F}[f_L] = F_L$? In other words, what is the function f_L which is band-limited to frequency L and whose Fourier transform agrees with that of F on $[-L, L]$. To answer this question set

$$G_L(\omega) = \begin{cases} 1, & |\omega| \leq L, \\ 0, & |\omega| > L. \end{cases}$$

Thus we can write

$$F_L(\omega) = F(\omega) G_L(\omega).$$

We write F_L in this form to stress the fact that we can obtain f_L by using the Convolution Theorem (Theorem 3.5) if we know the inverse Fourier transform of F and G_L. The first function f is given to us. In Example 3.5 we noted that for all $|x| \neq L$ we have

$$G_L(x) = \text{P.V.} \int_{-\infty}^{\infty} \frac{\sin \omega L}{\omega \pi} e^{i\omega x} \, d\omega.$$

Since G_L is an even function we can also write

$$G_L(x) = \text{P.V.} \int_{-\infty}^{\infty} \frac{\sin \omega L}{\omega \pi} e^{-i\omega x} \, d\omega.$$

We now exchange the roles of x and ω to obtain

$$G_L(\omega) = \text{P.V.} \frac{1}{2\pi} \int_{-\infty}^{\infty} \frac{2 \sin Lx}{x} e^{-i\omega x} \, dx.$$

Thus if $g_L(x) = \frac{2 \sin Lx}{x}$ then $\mathcal{F}[g_L](\omega) = G_L(\omega)$ in a certain sense (which is sufficient to obtain the desired result). From the Convolution Theorem

$$f_L = \frac{1}{2\pi}(f * g_L).$$

In other words

$$f_L(x) = \frac{1}{2\pi} \int_{-\infty}^{\infty} f(y) \frac{2 \sin[L(x-y)]}{x-y} \, dy$$

$$= \frac{1}{\pi} \int_{-\infty}^{\infty} \frac{f(y) \sin[L(x-y)]}{x-y} \, dy.$$

From the above formula we also obtain, as a corollary, that if $f \in G(\mathbb{R})$ and $\mathcal{F}[f](\omega) = 0$ for all $|\omega| \geq L$, then

$$f(x) = \frac{1}{\pi} \int_{-\infty}^{\infty} \frac{f(y) \sin[L(x-y)]}{x-y} \, dy.$$

This last formula is another characterization of functions which are band-limited to frequency at most L (see Exercise 4 of Section 5).

The Shannon Sampling Theorem

The Shannon Sampling Theorem provides us with a simple method of calculating any function which is band-limited to frequency at most L. This is an important and central result in signal processing.

Theorem 3.10: (**Shannon Sampling Theorem**) *If $f \in G(\mathbb{R})$ and $\mathcal{F}[f](\omega) = 0$ for all $|\omega| \geq L$, then*

$$f(x) = \sum_{n=-\infty}^{\infty} f(\tfrac{n\pi}{L}) \frac{\sin(Lx - n\pi)}{Lx - n\pi}.$$

"Proof": Let $F(\omega) = \mathcal{F}[f](\omega)$. From the Inverse Fourier Transform Theorem (which holds for our function)

$$f(x) = \text{P.V.} \int_{-\infty}^{\infty} F(\omega) e^{i\omega x}\, d\omega = \int_{-L}^{L} F(\omega) e^{i\omega x}\, d\omega.$$

Thus

$$f(\tfrac{n\pi}{L}) = \int_{-L}^{L} F(\omega) e^{\frac{i\omega n\pi}{L}}\, d\omega, \qquad n \in \mathbb{Z}.$$

Let us consider the complex Fourier series of F on $[-L, L]$ (see Section 2.3). It is given by

$$F(\omega) \sim \sum_{n=-\infty}^{\infty} c_n e^{-\frac{in\pi\omega}{L}}, \qquad |\omega| \leq L,$$

where

$$c_n = \frac{1}{2L} \int_{-L}^{L} F(\omega) e^{\frac{in\pi\omega}{L}}\, d\omega = \frac{1}{2L} f(\tfrac{n\pi}{L}), \qquad n \in \mathbb{Z}.$$

Define

$$H_L(\omega) = \sum_{n=-\infty}^{\infty} c_n e^{-\frac{in\pi\omega}{L}} = \frac{1}{2L} \sum_{n=-\infty}^{\infty} f(\tfrac{n\pi}{L}) e^{-\frac{in\pi\omega}{L}}.$$

Note that H_L is a $2L$-periodic function which "equals" F on the interval $[-L, L]$. (Recall that the series $\sum_{n=-\infty}^{\infty} c_n e^{-\frac{in\pi\omega}{L}}$ does not always equal $F(\omega)$ at every point of $[-L, L]$.) The function F is identically zero outside the interval $[-L, L]$. This being so, we can write

$$F(\omega) \text{ "=" } H_L(\omega) G_L(\omega)$$

where, as previously,

$$G_L(\omega) = \begin{cases} 1, & |\omega| \leq L, \\ 0, & |\omega| > L. \end{cases}$$

Chapter 3: The Fourier Transform

Thus

$$F(\omega) \text{``=''} \frac{1}{2L} \sum_{n=-\infty}^{\infty} f(\tfrac{n\pi}{L}) e^{-\frac{in\pi\omega}{L}} G_L(\omega).$$

We just proved that G_L is the Fourier transform of the function

$$g_L(x) = \frac{2 \sin Lx}{x}.$$

From the shift formula 3.4 it follows that $e^{-\frac{in\pi\omega}{L}} G_L(\omega)$ is the Fourier transform of the function

$$g_{L,n}(x) = \frac{2 \sin(Lx - n\pi)}{x - \frac{n\pi}{L}}.$$

Thus

$$F(\omega) \text{``=''} \frac{1}{2L} \sum_{n=-\infty}^{\infty} f(\tfrac{n\pi}{L}) \mathcal{F}[g_{L,n}](\omega).$$

Taking the inverse Fourier transforms of both sides we obtain

$$f(x) = \frac{1}{2L} \sum_{n=-\infty}^{\infty} f(\tfrac{n\pi}{L}) g_{L,n}(x) = \sum_{n=-\infty}^{\infty} f(\tfrac{n\pi}{L}) \frac{\sin(Lx - n\pi)}{Lx - n\pi}. \quad \blacksquare$$

Remarks: 1. The proof of Theorem 3.10 is somewhat "lacking" when we take the Fourier transform and its inverse. The proof with all the details would be rather technical and would not add much to our understanding of the theorem.

2. The importance of the Shannon Sampling Theorem may be stated as follows. Functions which are band-limited to frequency at most L are totally determined by their values at a sequence of evenly spaced points with distance $\frac{\pi}{L}$ between consecutive points. In addition, the theorem gives us an exact formula for recovering such functions at every point x based only on their values at that sequence.

3. The function $\frac{\sin(Lx-n\pi)}{Lx-n\pi}$ is equal 1 at $x = \frac{n\pi}{L}$ (this is a removable point of discontinuity), and vanishes at each of the points $x = \frac{k\pi}{L}$, $k \in \mathbb{Z}$, $k \neq n$. Thus for each $n \in \mathbb{Z}$ we have a simple formal equality between both sides of the formula at the points $x = \frac{n\pi}{L}$.

4. If we take any $f \in G(\mathbb{R})$, and set

$$g(x) = \sum_{n=-\infty}^{\infty} f(\tfrac{n\pi}{L}) \frac{\sin(Lx - n\pi)}{Lx - n\pi},$$

then we obtain a function $g \in G(\mathbb{R})$ which is band-limited to frequency at most L, and such that g agrees with f at the points $x = \frac{n\pi}{L}$, $n \in \mathbb{Z}$.

5. The functions $\left\{ \frac{\sin(Lx-n\pi)}{Lx-n\pi} \right\}_{n=-\infty}^{\infty}$ are an orthonormal system with respect to the inner product

$$\langle f, g \rangle = \frac{L}{\pi} \int_{-\infty}^{\infty} f(x) \overline{g(x)} \, dx.$$

Time- and Band-Limited Functions

There do not exist functions (signals) in $G(\mathbb{R})$ (other than the identically zero signal) which are both time- and band-limited. We will prove this rather surprising fact by two distinct methods of proof. The first method is an application of the Shannon Sampling Theorem. The second method is based on the theory of analytic functions. We prove that the Fourier transform of any time-limited function may be extended to the complex plane as an analytic function. Since the only analytic function which vanishes on an interval is the identically zero function, the desired result then follows.

Method 1. Assume f is band-limited to $[-L, L]$, i.e., $F(\omega) = 0$ for $|\omega| > L$, and time-limited to $[-M, M]$, i.e., $f(x) = 0$ for $|x| > M$. From the Shannon Sampling Theorem (Theorem 3.10) and since f is time-limited to $[-M, M]$

$$f(x) = \sum_{n=-\infty}^{\infty} f(\tfrac{n\pi}{L}) \frac{\sin(Lx - n\pi)}{Lx - n\pi}$$

$$= \sum_{n=-K}^{K} f(\tfrac{n\pi}{L}) \frac{\sin(Lx - n\pi)}{Lx - n\pi}$$

$$= \sin Lx \sum_{n=-K}^{K} \frac{f(\tfrac{n\pi}{L}) \cdot (-1)^n}{Lx - n\pi}$$

where K is any integer greater than $\frac{LM}{\pi} + 1$. For any constants $\{a_n\}_{n=-K}^{K}$

$$\sum_{n=-K}^{K} \frac{a_n}{Lx - n\pi} = \frac{p(x)}{q(x)}$$

where $q(x) = \prod_{n=-K}^{K}(Lx - n\pi)$ (the common denominator) and $p(x)$ is an algebraic polynomial of degree at most $2K$. Thus

$$f(x) = \sin Lx \cdot \frac{p(x)}{q(x)}$$

for some p as above. Since f vanishes identically off $[-M, M]$, it follows that p vanishes identically off $[-M, M]$. But p is a polynomial and thus p is identically zero on all of $\mathbb{R}$. This implies in turn that f is identically zero.

Method 2. We start by extending our definition of the Fourier transform (but not of our function f) to the complex plane $\mathbb{C}$. That is, for $\xi = \omega + is$, let

$$F(\xi) = \frac{1}{2\pi} \int_{-\infty}^{\infty} f(x) e^{-i\xi x} dx$$

if the integral exists. Note that

$$F(\xi) = \frac{1}{2\pi} \int_{-\infty}^{\infty} f(x) e^{sx} e^{-i\omega x} dx.$$

This is, for each fixed s, the Fourier transform of $f(x)e^{sx}$. This integral will certainly exist and satisfy the conditions of Theorem 3.1 with respect to ω if $f(x)e^{sx} \in G(\mathbb{R})$. In fact more may be true.

Theorem 3.11: *Assume $f(x)e^{s_1 x}, f(x)e^{s_2 x} \in G(\mathbb{R})$ where $s_1 < s_2$. Then the function*

$$F(\xi) = \frac{1}{2\pi} \int_{-\infty}^{\infty} f(x) e^{-i\xi x} dx$$

exists and is analytic in the strip

$$\mathcal{D} = \{\xi \mid \xi = \omega + is, \quad -\infty < \omega < \infty, \quad s_1 < s < s_2\}.$$

Proof: We divide the proof into three steps. We first prove that $F(\xi)$ exists for each $\xi \in \mathcal{D}$. We then prove that F is continuous on $\mathcal{D}$, and we finally prove that it is in fact analytic thereon.

Given $\xi = \omega + is$ where $s_1 < s < s_2$, we must prove that $f(x)e^{sx}$ is absolutely integrable in order to show that $F(\xi)$ exists. For $x \geq 0$,

$$|f(x)e^{sx}| = |f(x)|e^{sx} \leq |f(x)|e^{s_2 x} = |f(x)e^{s_2 x}|.$$

For $x \leq 0$,

$$|f(x)e^{sx}| = |f(x)|e^{sx} \leq |f(x)|e^{s_1 x} = |f(x)e^{s_1 x}|.$$

Thus

$$\int_{-\infty}^{\infty} |f(x)e^{sx}| \, dx \leq \int_{0}^{\infty} |f(x)e^{s_2 x}| \, dx + \int_{-\infty}^{0} |f(x)e^{s_1 x}| \, dx < \infty$$

since $f(x)e^{s_1 x}$ and $f(x)e^{s_2 x}$ are themselves absolutely integrable on all of $\mathbb{R}$.

We now prove that $F(\xi)$ is continuous on $\mathcal{D}$. To this end let ξ_n be a sequence in this domain which converges to ξ^*, where $\xi^* \in \mathcal{D}$. We have to prove that $\lim_{n\to\infty} F(\xi_n) = F(\xi^*)$. By definition

$$|F(\xi_n) - F(\xi^*)| = \frac{1}{2\pi} \left| \int_{-\infty}^{\infty} \left(e^{-i\xi_n x} - e^{-i\xi^* x} \right) f(x) \, dx \right|$$

$$\leq \frac{1}{2\pi} \int_{-\infty}^{\infty} \left| e^{-i\xi_n x} - e^{-i\xi^* x} \right| |f(x)| \, dx.$$

Clearly

$$\lim_{n \to \infty} \left| e^{-i\xi_n x} - e^{-i\xi^* x} \right| = 0$$

for each $x \in \mathbb{R}$. If we can find an absolutely integrable function g for which

(3.2) $$\left|e^{-i\xi_n x} - e^{-i\xi^* x}\right| \cdot |f(x)| \leq g(x)$$

then from the Lebesgue Dominated Convergence Theorem (Theorem 3.2) it will follow that

$$\lim_{n \to \infty} \int_{-\infty}^{\infty} \left|e^{-i\xi_n x} - e^{-i\xi^* x}\right| \cdot |f(x)|\, dx = 0$$

and thus $\lim_{n \to \infty}[F(\xi_n) - F(\xi^*)] = 0$. We know that $s_1 < \operatorname{Re}(-i\xi_n), \operatorname{Re}(-i\xi^*) < s_2$, and therefore

$$\left|e^{-i\xi_n x} - e^{-i\xi^* x}\right| \leq |e^{s_2 x}| + |e^{s_2 x}| = 2e^{s_2 x}$$

for $x \geq 0$, while

$$\left|e^{-i\xi_n x} - e^{-i\xi^* x}\right| \leq 2e^{s_1 x}$$

for $x \leq 0$. The function

$$g(x) = \begin{cases} 2e^{s_2 x}|f(x)|, & x \geq 0, \\ 2e^{s_1 x}|f(x)|, & x < 0, \end{cases}$$

is absolutely integrable and satisfies (3.2). Hence F is continuous on the complex strip $\mathcal{D}$.

We now prove that F is analytic in the domain $\mathcal{D}$.

Let γ be a simple and closed path inside $\mathcal{D}$. By the definition of F

$$\oint_\gamma F(\xi)\, d\xi = \oint_\gamma \left(\int_{-\infty}^{\infty} e^{-i\xi x} f(x)\, dx\right) d\xi.$$

Since $e^{-i\xi x} f(x)$ is absolutely integrable, we can apply Fubini's Theorem (Theorem B.4) to exchange the order of integration to obtain

$$\oint_\gamma F(\xi)\, d\xi = \int_{-\infty}^{\infty} \left(\oint_\gamma e^{-i\xi x} d\xi\right) f(x)\, dx.$$

The function $G(\xi) = e^{-i\xi x}$ (where x is constant) is an analytic function of the variable ξ defined on all of $\mathbb{C}$. Hence from Cauchy's Theorem (Theorem A.1)

$$\oint_\gamma e^{-i\xi x} d\xi = 0$$

for each x. Thus

$$\oint_\gamma F(\xi)\, d\xi = 0.$$

From the fact that F is continuous on $\mathcal{D}$ and Morera's Theorem (Theorem A.2) it follows that F is analytic on $\mathcal{D}$. ∎

Remark: Theorem 3.11 may be proven without recourse to the theorems of Cauchy and Morera. The function F is analytic in the strip $\mathcal{D}$ if its derivative exists at every point of $\mathcal{D}$. It is possible, via a somewhat more elaborate application of the Lebesgue Dominated Convergence Theorem (Theorem 3.2) to directly prove that $F'(\xi)$ exists at every point $\xi \in \mathcal{D}$.

Why is Theorem 3.11 relevant to us here? Assume f is time-limited and piecewise continuous. Then for each and every s, $f(x)e^{sx}$ is time-limited and piecewise continuous and hence in $G(\mathbb{R})$. Thus, from Theorem 3.11, the "extended Fourier transform" F is analytic on all of $\mathbb{C}$. Restricted to the real axis $\xi = \omega$, this is the usual Fourier transform. If f is also band-limited, then f vanishes identically on $(-\infty, -L) \cup (L, \infty)$ for some constant L. But the only analytic function which vanishes thereon is the identically zero function. Thus F is identically zero, which in turn implies that f is identically zero.

Review Exercises

1. Calculate the Fourier transform of

$$f(x) = \begin{cases} 1 - x^2, & |x| \le 1, \\ 0, & |x| > 1, \end{cases}$$

   and show that

$$\int_0^\infty \left(\frac{x \cos x - \sin x}{x^3} \right) \cos \frac{x}{2} \, dx = \frac{3\pi}{16}.$$

2. For each $x > 0$, we define $f(x) = e^{-x} \cos x$. Let $\tilde{f}$ be the odd continuation of f. Prove that for all $x \ne 0$,

$$\frac{2}{\pi} \int_0^\infty \frac{t^3 \sin xt}{t^4 + 4} \, dt = \tilde{f}(x).$$

3. Let

$$f(x) = \begin{cases} e^{-x}, & x > 0, \\ 0, & x \le 0. \end{cases}$$

   (a) Calculate the Fourier transform F of f.
   (b) Determine $f * f$ and $(f * f) * (f * f)$.
   (c) Find $\mathcal{F}[(f * f) * (f * f)]$.
   (d) Calculate the integral $\int_{-\infty}^\infty \frac{1}{(1+x^2)^4} \, dx$.

4. Let
$$f(x) = \begin{cases} 1, & |x| \le 1, \\ 0, & |x| > 1. \end{cases}$$

(a) Determine $f * f$.
(b) Calculate the integrals $\int_{-\infty}^{\infty} \frac{\sin^2 x}{x^2} dx$ and $\int_{-\infty}^{\infty} \frac{\sin^4 x}{x^4} dx$.

5. For each $a > 0$, let $f_a(x) = e^{-a|x|}$, and $g_a(x) = \frac{2a}{x^2+a^2}$.
   (a) Find the Fourier transform of f_a.
   (b) Find the Fourier transform of g_a.
   (c) Does there exist a function $\varphi \in G(\mathbb{R})$ such that $\int_{-\infty}^{\infty} \frac{\varphi(t)}{(x-t)^2+16} dt = \frac{1}{x^2+49}$? If yes, find it. If no, explain why it cannot exist.
   (d) Does there exist a function $\varphi \in G(\mathbb{R})$ such that $\int_{-\infty}^{\infty} \frac{\varphi(t)}{(x-t)^2+49} dt = \frac{1}{x^2+16}$? If yes, find it. If no, explain why it cannot exist.

6. Let
$$f(x) = \begin{cases} x - a, & a < x < 4a, \\ 0, & \text{otherwise}, \end{cases}$$

where $a > 0$, and let $F(\omega)$ be the Fourier transform of f. Let $g \in G(\mathbb{R})$ be a continuous function whose Fourier transform $G(\omega)$ satisfies the equality $G(\omega) = 2\pi F(\omega) F(-\omega)$. Calculate $g(2a)$.

7. Prove that for every $a > b > 0$ the following inequality holds:
$$\int_0^{\infty} \frac{\cos ax}{x^2+b^2} dx > \int_0^{\infty} \frac{\cos bx}{x^2+a^2} dx.$$

8. Solve the following boundary-value problem:
$$\begin{cases} u_t = 4u_{xx}, & -\infty < x < \infty, \quad 0 < t < \infty, \\ u(x,0) = f(x), & -\infty < x < \infty, \end{cases}$$

where
$$f(x) = \begin{cases} 1, & |x| \le 2, \\ 0, & |x| > 2. \end{cases}$$

9. (B. Paneah) Assume $f \in G(\mathbb{R})$, $f(x) = 0$ for $|x| \ge \frac{1}{2}$, and f is real-valued on $\mathbb{R}$. Prove that
$$\int_{-\infty}^{\infty} |\mathcal{F}[f](\omega)|^2 T(\omega) \, d\omega = 0$$
for every trigonometric polynomial T.

10. Find all values ω for which $\int_{-\infty}^{\infty} \frac{\sin^2 5x}{x^2} \cos \omega x \, dx = 0$.

11. Let $f \in C(\mathbb{R}^n)$ be absolutely integrable on $\mathbb{R}^n$. For real $\omega_1, \omega_2, \ldots, \omega_n$ we define

$$\mathcal{F}[f](\omega_1, \ldots, \omega_n) =$$

$$\frac{1}{(2\pi)^n} \underbrace{\int_{-\pi}^{\pi} \cdots \int_{-\pi}^{\pi}}_{n \text{ times}} f(x_1, \ldots, x_n) e^{-i(\omega_1 x_1 + \cdots + \omega_n x_n)} \, dx_1 \cdots dx_n.$$

The function $\mathcal{F}[f]$ (defined on $\mathbb{R}^n$) is said to be the *multivariate Fourier transform of f*. Show that if $f_1, f_2, \ldots, f_n$ are univariate continuous functions in $G(\mathbb{R})$, and if

$$f(x_1, x_2, \ldots, x_n) = f_1(x_1) f_2(x_2) \cdots f_n(x_n)$$

then

$$\mathcal{F}[f](\omega_1, \ldots, \omega_n) = \mathcal{F}[f_1](\omega_1) \mathcal{F}[f_2](\omega_2) \cdots \mathcal{F}[f_n](\omega_n).$$

Chapter 4
The Laplace Transform

0. Introduction

The theory of the Laplace transform has a long history involving many mathematicians who have significantly contributed to its development. Among them are to be found Euler, Lagrange, Laplace, Poincaré, and Doetsch. The Laplace transform has found use both in mathematics and in its diverse applications. The development of the Laplace transform is intimately connected with methods for solving differential equations. This chapter is an introduction to the theory and applications of the Laplace transform. In Sections 1, 2, 3, 5, 7, and 8, we present the basic theory and some of its common applications. Section 4 discusses two important "functions" and applications of the Laplace transform in connection with these functions. In Section 6 we present some applications of the Laplace transform to the computing of integrals.

1. Definition and Examples

Let f be a complex-valued piecewise continuous function on $[0, \infty)$. For every $s \in \mathbb{R}$ define

$$\mathcal{L}[f](s) = \int_0^\infty e^{-st} f(t)\, dt$$

if this integral converges (exists). This defines a function $\mathcal{L}[f](s)$ which we call the *Laplace transform* of f. As we shall see in the examples and theorems of this chapter, the Laplace transform of f often exists for some values of s but does not exist for other values of s.

It will sometimes be desirable to think of f as being defined on all of $\mathbb{R}$. In such cases we assume that $f(t) = 0$ for every $t < 0$. This assumption allows us to present a simple connection between the Laplace transform of f and its Fourier transform, namely,

$$\mathcal{L}[f](s) = \int_0^\infty e^{-st} f(t)\, dt = \int_{-\infty}^\infty e^{-i(-is)t} f(t)\, dt = 2\pi \mathcal{F}[f](-is).$$

Chapter 4: The Laplace Transform

Recall that in our discussion of the Fourier transform we always considered only real ω. We return to this connection in Section 7.

Before discussing in detail the Laplace transform and some of its properties, let us first present a few examples.

Example 4.1: Set $f(t) = 1$ for every $t \geq 0$. Then

$$\mathcal{L}[f](s) = \int_0^\infty e^{-st} dt = \lim_{L \to \infty} \left(\frac{e^{-st}}{-s} \bigg|_0^L \right) = \frac{1}{s}, \quad s > 0.$$

For $s \leq 0$ the integral does not converge, and therefore $\mathcal{L}[f](s) = \frac{1}{s}$ only for $s > 0$. In general we will also specify the domain whereon $\mathcal{L}[f]$ is defined.

Example 4.2: Set $f(t) = e^{at}$, where $a \in \mathbb{R}$. Then

$$\mathcal{L}\left[e^{at}\right](s) = \int_0^\infty e^{-st} e^{at} dt = \int_0^\infty e^{-(s-a)t} dt = \frac{1}{s-a}, \quad s > a.$$

Example 4.3: Set $f(t) = e^{zt}$, where $z = x + iy \in \mathbb{C}$. Then

$$\mathcal{L}\left[e^{zt}\right] = \int_0^\infty e^{-st} e^{(x+iy)t} dt = \int_0^\infty e^{-(s-x-iy)t} dt = \frac{1}{s-z}, \quad s > x.$$

Clearly Example 4.2 is a particular case of Example 4.3.

The Laplace transform is a linear operation on a set of functions, i.e.,

$$\mathcal{L}[af + bg](s) = a\mathcal{L}[f](s) + b\mathcal{L}[g](s)$$

for functions f and g, and $a, b \in \mathbb{C}$ (for which both sides of the equation are defined). This property of linearity, often unstated, is fundamental and important.

Example 4.4: Let $f(t) = \sin at$, where $a \in \mathbb{R}$. It is possible to compute $\mathcal{L}[\sin at](s)$ from its definition (twice integrating by parts). However, we will compute it by using the linearity property and the result of Example 4.3. We start with the identity

$$\sin at = \frac{e^{iat} - e^{-iat}}{2i}.$$

Thus

$$\mathcal{L}[\sin at](s) = \mathcal{L}\left[\frac{e^{iat} - e^{-iat}}{2i}\right](s) = \frac{1}{2i} \mathcal{L}\left[e^{iat}\right](s) - \frac{1}{2i} \mathcal{L}\left[e^{-iat}\right](s).$$

From Example 4.3 we obtain

$$\mathcal{L}[\sin at](s) = \frac{1}{2i} \left[\frac{1}{s - ia} - \frac{1}{s + ia} \right] = \frac{a}{s^2 + a^2}, \quad s > 0.$$

Example 4.5: Set $f(t) = \cos at$, where $a \in \mathbb{R}$. In the same way as in the previous example we obtain

$$\mathcal{L}[\cos at](s) = \frac{s}{s^2 + a^2}, \qquad s > 0.$$

In all these examples we see that the Laplace transform is defined for every value of s greater than some constant. That is, the domain of definition of $\mathcal{L}[f](s)$ is in general of the form (c, ∞) or $[c, \infty)$, for some constant c (which of course depends on f). The reason for this is that if $s_1 < s_2$ then $e^{-s_1 t} > e^{-s_2 t} > 0$, for every $t > 0$. Hence if the integral converges at s_1, it will almost surely converge for s_2. We formally state

Proposition 4.1: *Let f be a complex-valued piecewise continuous function defined on the interval $[0, \infty)$. If there exist real constants a and K such that*

$$|f(t)| \leq K e^{at}, \qquad t \geq 0,$$

then $\mathcal{L}[f](s)$ is defined for every $s > a$.

Proof: From the definition

$$\mathcal{L}[f](s) = \int_0^\infty e^{-st} f(t)\, dt.$$

The function $e^{-st} f(t)$ is piecewise continuous on $[0, \infty)$. We prove that it is absolutely integrable for each $s > a$. If $s > a$, then

$$\int_0^\infty \left| e^{-st} f(t) \right| dt = \int_0^\infty e^{-st} |f(t)|\, dt \leq \int_0^\infty e^{-st} K e^{at}\, dt = \frac{K}{s-a}. \qquad \blacksquare$$

Examples 4.1–4.5 are testimony to the accuracy of this theorem.

Exercises

1. Does the function $f(t) = e^{t^2}$ have a Laplace transform? If yes, find it and its domain of definition. If not, explain why.

2. Find the Laplace transform of

$$f(t) = \begin{cases} \sin t, & 0 \leq t \leq 2\pi, \\ 0, & 2\pi < t. \end{cases}$$

3. Prove that if $f : [0, \infty) \to \mathbb{C}$ is piecewise continuous and p-periodic $(p > 0)$, then

$$\mathcal{L}[f](s) = \frac{1}{1 - e^{-ps}} \int_0^p e^{-st} f(t)\, dt, \qquad s > 0.$$

4. Let $f : [0, \infty) \to \mathbb{R}$ be a 4-periodic function such that

$$f(t) = \begin{cases} 1, & 0 \leq t < 2, \\ -1, & 2 \leq t < 4. \end{cases}$$

Prove that

$$\mathcal{L}[f](s) = \frac{\tanh s}{s}, \qquad s > 0.$$

2. More Formulae and Examples

We present in this section a number of useful formulae which will help us to quickly and efficiently compute the Laplace transform of many different functions. These formulae should be compared with the formulae for the Fourier transform.

Theorem 4.2: (**Derivative Formula**) *Let f be a continuous function on the interval $[0, \infty)$ and f' be piecewise continuous thereon. Assume there exist constants a and K such that*

$$|f(t)| \leq Ke^{at}, \qquad t \geq 0.$$

Then the Laplace transform of f', $\mathcal{L}[f'](s)$, is defined for every $s > a$, and

(4.1) $$\boxed{\mathcal{L}[f'](s) = s\mathcal{L}[f] - f(0)}$$

Proof: For every choice of $s \in \mathbb{R}$, the function $e^{-st}f'(t)$ is piecewise continuous on every finite interval $[0, L]$. Let

$$0 < x_1 < x_2 < \cdots < x_{n-1} < L$$

be the points of discontinuity of $e^{-st}f'(t)$ in the open interval $(0, L)$. Set $x_0 = 0$ and $x_n = L$. Then for every i, $1 \leq i \leq n$, $e^{-st}f'(t)$ is continuous on (x_{i-1}, x_i),

and therefore

$$\int_0^L e^{-st} f'(t)\, dt = \sum_{i=1}^n \int_{x_{i-1}}^{x_i} e^{-st} f'(t)\, dt$$

$$= \sum_{i=1}^n \left[e^{-st} f(t) \Big|_{x_{i-1}}^{x_i} + s \int_{x_{i-1}}^{x_i} e^{-st} f(t)\, dt \right]$$

$$= \sum_{i=1}^n \left[e^{-sx_i} f(x_i) - e^{-sx_{i-1}} f(x_{i-1}) \right] + s \sum_{i=1}^n \int_{x_{i-1}}^{x_i} e^{-st} f(t)\, dt$$

$$= e^{-sL} f(L) - f(0) + s \int_0^L e^{-st} f(t)\, dt.$$

By Proposition 4.1, $\mathcal{L}[f](s)$ is defined for each $s > a$. Clearly

$$\mathcal{L}[f](s) = \lim_{L \to \infty} \int_0^L e^{-st} f(t)\, dt, \qquad s > a.$$

Furthermore

$$\lim_{L \to \infty} \left| e^{-sL} f(L) \right| \le \lim_{L \to \infty} e^{-sL} K e^{aL} = K \lim_{L \to \infty} e^{-(s-a)L} = 0, \qquad s > a.$$

Hence

$$\mathcal{L}[f'](s) = \lim_{L \to \infty} \int_0^L e^{-st} f'(t)\, dt$$

$$= \lim_{L \to \infty} \left[e^{-sL} f(L) - f(0) + s \int_0^L e^{-st} f(t)\, dt \right]$$

$$= s \mathcal{L}[f](s) - f(0)$$

for every $s > a$. ∎

The immediate generalization of this derivative formula is presented in the next theorem.

Theorem 4.3: *Let $f, f', f'', \ldots, f^{(n-1)}$ be continuous functions on the interval $[0, \infty)$. Let $f^{(n)}$ be piecewise continuous thereon, and assume there exist constants a and K such that*

$$\left| f^{(j)}(t) \right| \le K e^{at}, \qquad t \ge 0, \quad j = 0, 1, 2, \ldots, n-1.$$

Then $\mathcal{L}\left[f^{(n)}\right](s)$ is defined for every $s > a$ and

(4.2) $$\boxed{\mathcal{L}\left[f^{(n)}\right](s) = s^n \mathcal{L}[f](s) - s^{n-1} f(0) - s^{n-2} f'(0) - \cdots - f^{(n-1)}(0)}$$

Proof: The case $n = 2$ is obtained by twice applying Theorem 4.2. That is,

$$\mathcal{L}[f''](s) = s\mathcal{L}[f'](s) - f'(0)$$
$$= s\left[s\mathcal{L}[f](s) - f(0)\right] - f'(0)$$
$$= s^2\mathcal{L}[f](s) - sf(0) - f'(0).$$

The full proof is easily obtained by a simple induction on n. ∎

A formula which is, in some sense, similar to (4.2) is

(4.3) $$\boxed{\mathcal{L}[t^n f(t)](s) = (-1)^n \frac{d^n}{ds^n} \mathcal{L}[f](s)}$$

This formula holds under conditions similar to those in Theorems 4.2 and 4.3 by Leibniz's rule (since $|f(t)| \leq Ke^{at}$ implies $|tf(t)| \leq K_b e^{bt}$ for some constant K_b and any $b > a$). It will suffice to prove this formula for $n = 1$ since for every $n > 1$ the formula is obtained by n iterations of this case. We must prove that

$$\mathcal{L}[tf(t)](s) = -\frac{d}{ds}\mathcal{L}[f](s).$$

Proof: Starting with the right-hand side we have

$$-\frac{d}{ds}\mathcal{L}[f](s) = -\frac{d}{ds}\int_0^\infty e^{-st} f(t)\, dt$$
$$= -\int_0^\infty -te^{-st} f(t)\, dt$$
$$= \int_0^\infty te^{-st} f(t)\, dt$$
$$= \mathcal{L}[tf(t)](s).$$
∎

(4.4) $$\boxed{\mathcal{L}[e^{at} f(t)](s) = \mathcal{L}[f](s - a), \quad a \in \mathbb{R}}$$

Proof:

$$\mathcal{L}\left[e^{at} f(t)\right](s) = \int_0^\infty e^{-st} e^{at} f(t)\, dt = \int_0^\infty e^{-(s-a)t} f(t)\, dt = \mathcal{L}[f](s - a).$$
∎

(4.5) $$\boxed{\mathcal{L}[f(at)](s) = \tfrac{1}{a}\mathcal{L}[f](\tfrac{s}{a}), \qquad a > 0}$$

Proof:
$$\mathcal{L}[f(at)](s) = \int_0^\infty e^{-st} f(at)\, dt.$$
Substitute $x = at$ to obtain
$$\mathcal{L}[f(at)](s) = \int_0^\infty e^{-s\frac{x}{a}} f(x) \frac{dx}{a} = \frac{1}{a}\int_0^\infty e^{-(\frac{s}{a})x} f(x)\, dx = \frac{1}{a}\mathcal{L}[f](\frac{s}{a}). \quad \blacksquare$$

All the formulae so far obtained are simple but useful. We can, for example, use them to more immediately obtain Examples 4.1, 4.2, 4.4, and 4.5.

Example 4.6: Let us calculate the Laplace transform of $f(t) = t^n$, $n \in \mathbb{Z}_+$, by two different methods.

Method 1: From Example 4.1 and (4.3),
$$\mathcal{L}[t^n \cdot 1](s) = (-1)^n \frac{d^n}{ds^n} \mathcal{L}[1](s) = (-1)^n \frac{d^n}{ds^n} \frac{1}{s}, \qquad s > 0.$$
It is easily verified that
$$\frac{d^n}{ds^n} \frac{1}{s} = (-1)^n \frac{n!}{s^{n+1}}$$
and thus
$$\mathcal{L}[t^n](s) = \frac{n!}{s^{n+1}}, \qquad s > 0.$$

Method 2: Clearly $f^{(n)}(t) = n!$, and therefore using Example 4.1 and (4.2) we have
$$\frac{n!}{s} = \mathcal{L}[n!](s) = \mathcal{L}\left[f^{(n)}\right](s)$$
$$= s^n \mathcal{L}[f](s) - s^{n-1} f(0) - s^{n-2} f'(0) - \cdots - f^{(n-1)}(0).$$
For $f(t) = t^n$ we have
$$f(0) = f'(0) = \cdots = f^{(n-1)}(0) = 0.$$
Hence
$$\frac{n!}{s} = s^n \mathcal{L}[t^n](s)$$
and therefore
$$\mathcal{L}[t^n](s) = \frac{n!}{s^{n+1}}, \qquad s > 0.$$

Table of Laplace Transforms

	Function	Laplace Transform
1	1	$\frac{1}{s}, \quad s>0$
2	$e^{at}, \quad a \in \mathbb{R}$	$\frac{1}{s-a}, \quad s>a$
3	$e^{zt}, \quad z \in \mathbb{C}$	$\frac{1}{s-z}, \quad s>\operatorname{Re}(z)$
4	$\sin at, \quad a \in \mathbb{R}$	$\frac{a}{s^2+a^2}, \quad s>0$
5	$\cos at, \quad a \in \mathbb{R}$	$\frac{s}{s^2+a^2}, \quad s>0$
6	$t^n, \quad n \in \mathbb{Z}_+$	$\frac{n!}{s^{n+1}}, \quad s>0$
7	$e^{at}\sin bt, \quad a,b \in \mathbb{R}$	$\frac{b}{(s-a)^2+b^2}, \quad s>a$
8	$e^{at}\cos bt, \quad a,b \in \mathbb{R}$	$\frac{s-a}{(s-a)^2+b^2}, \quad s>a$
9	$u_c(t), \quad c>0$	$\frac{e^{-cs}}{s}, \quad s>0$
10	$t^n e^{at}, \quad a \in \mathbb{R}, n \in \mathbb{Z}_+$	$\frac{n!}{(s-a)^{n+1}}, \quad s>a$
11	$f'(t)$	$s\mathcal{L}[f](s) - f(0)$
12	$f^{(n)}(t)$	$s^n\mathcal{L}[f](s) - s^{n-1}f(0) - \cdots - f^{(n-1)}(0)$
13	$t^n f(t)$	$(-1)^n \frac{d^n}{ds^n}\mathcal{L}[f](s)$
14	$e^{at}f(t), \quad a \in \mathbb{R}$	$\mathcal{L}[f](s-a)$
15	$f(at), \quad a>0$	$\frac{1}{a}\mathcal{L}[f]\left(\frac{s}{a}\right)$

Example 4.7: Let $f(t) = e^{at}\sin bt$, where $a, b \in \mathbb{R}$. We use Example 4.4 and (4.4) to calculate $\mathcal{L}[f(t)](s)$.

$$\mathcal{L}\left[e^{at}\sin bt\right](s) = \mathcal{L}[\sin bt](s-a) = \frac{b}{(s-a)^2+b^2}, \qquad s > a.$$

Example 4.8: Similarly for $f(t) = e^{at}\cos bt$,

$$\mathcal{L}\left[e^{at}\cos bt\right](s) = \mathcal{L}[\cos bt](s-a) = \frac{s-a}{(s-a)^2+b^2}, \qquad s > a.$$

Let $a \in \mathbb{R}$ and $F(s)$ be a function which is defined for each $s > a$. We may ask if there exists a function $f(t)$ for which $\mathcal{L}[f](s) = F(s)$. The function $f(t)$, if it exists, is called the *inverse Laplace transform* of $F(s)$. In general, the answer to this question is in the negative. Necessary and sufficient conditions on $F(s)$ in order that it have an inverse Laplace transform are not simple. We conclude this section by stating one of the simplest necessary conditions.

Proposition 4.4: *Let f be a complex-valued piecewise continuous function on $[0, \infty)$. If there exist constants K and a such that*

$$|f(t)| \leq Ke^{at}, \qquad t \geq 0,$$

then

$$\lim_{s \to \infty} \mathcal{L}[f](s) = 0.$$

Proof: For every $s > a$,

$$|\mathcal{L}[f](s)| = \left|\int_0^\infty e^{-st}f(t)\,dt\right| \leq \int_0^\infty e^{-st}|f(t)|\,dt$$

$$\leq \int_0^\infty e^{-st}Ke^{at}dt = \frac{K}{s-a}$$

and therefore $\lim_{s \to \infty} \mathcal{L}[f](s) = 0$. ∎

For the convenience of the reader we have summarized all the formulae and examples of this and the previous section in one table. We will sometimes refer to a certain formula or example by its number in the table.

Exercises

1. Calculate the Laplace transform of each of the following functions:

   (a) $e^{-t}\cos 2t$

   (b) $e^{-4t}\cosh 2t$

   (c) $(t^2+1)^2$

   (d) $3\cosh t - 4\sinh 5t$

   (e) $t^n \sin t$

2. Calculate the Laplace transform of $f(t) = \frac{\sin t}{t}$. (Hint: First compute $\frac{d}{ds}\mathcal{L}[f](s)$.)

3. Find f if it is known that $\mathcal{L}[f](s) = \frac{d^n}{ds^n}\left[\frac{1}{s^2 - a^2}\right]$, $a > 0$.

4. Prove each of the following formulae:

    (a) $\mathcal{L}\left[\sin^2 t\right](s) = \dfrac{2}{s(s^2 + 4)}$

    (b) $\mathcal{L}[A\cos(\omega t + \theta)](s) = \dfrac{A(s\cos\theta - \omega\sin\theta)}{s^2 + \omega^2}$

    (c) $\mathcal{L}[\cos at \cosh at](s) = \dfrac{s^3}{s^4 + 4a^4}$

    (d) $\mathcal{L}\left[(t^2 - 5t + 6)e^{2t}\right] = \dfrac{6s^2 - 29s + 36}{(s - 2)^3}$

5. By using the definition and properties of the Laplace transform, compute each of the following integrals:

    (a) $\displaystyle\int_0^\infty t e^{-2t} \cos t \, dt$

    (b) $\displaystyle\int_0^\infty t^3 e^{-t} \sin t \, dt$

    (c) $\displaystyle\int_0^\infty x^4 e^{-x} \, dx$

    (d) $\displaystyle\int_0^\infty x^6 e^{-3x} \, dx$

3. Applications to Ordinary Differential Equations

This section is about applying the Laplace transform to solve second order linear homogeneous constant coefficient ordinary differential equations. The unknown function in all these equations is assumed to be defined on the full interval $[0, \infty)$. The general form of such equations is

$$\begin{cases} ay''(t) + by'(t) + cy(t) = 0, & t > 0, \\ y(0) = y_0, \quad y'(0) = y_1, \end{cases}$$

where a, b, c, y_0, and y_1, are given real constants. We will describe a general method of finding the particular solution $y(t)$. The reader has surely met other methods of solving such equations in courses on ordinary differential equations. It is not our aim to simply present yet one more method of solving this equation. We are interested in showing how the Laplace transform can be used to solve equations of this type, and this is a prototype. The use of the Laplace transform is not limited to solving only these particular equations. It may be applied to equations of higher order, to systems of equations, and more.

We first use this method to solve three typical examples which correspond to the possible root behaviour of the characteristic polynomial. At the end of the section we present a general method for the solution of all such equations.

Example 4.9: We solve the initial value problem

$$\begin{cases} y''(t) - y'(t) - 6y(t) = 0, & t > 0, \\ y(0) = 1, \quad y'(0) = -1. \end{cases}$$

Solution: Apply the Laplace transform to both sides of the equation to obtain

$$0 = \mathcal{L}[0] = \mathcal{L}[y'' - y' - 6y].$$

From the linearity of the Laplace transform we have

$$\mathcal{L}[y'' - y' - 6y] = \mathcal{L}[y''] - \mathcal{L}[y'] - 6\mathcal{L}[y].$$

We now use the derivative formula (see Formulae 11 and 12 in the table):

$$\mathcal{L}[y'] = s\mathcal{L}[y] - y(0),$$
$$\mathcal{L}[y''] = s^2\mathcal{L}[y] - sy(0) - y'(0).$$

From the initial conditions, we see that

$$(s^2 - s - 6)\mathcal{L}[y] - s + 2 = 0$$

and thus

$$\mathcal{L}[y] = \frac{s-2}{s^2 - s - 6}.$$

This finishes the first step in our method. We have transformed our differential equation in y to an easily solved algebraic equation in $\mathcal{L}[y]$. It now remains to determine y from $\mathcal{L}[y]$.

Let us denote by $\mathcal{L}^{-1}$ the inverse Laplace transform operation. At this stage we can only say that $\mathcal{L}^{-1}$ exists. Since $\mathcal{L}$ is a linear operation, it follows that its inverse $\mathcal{L}^{-1}$ is also a linear operation. That is to say,

$$\mathcal{L}^{-1}[aF + bG](t) = a\mathcal{L}^{-1}[F](t) + b\mathcal{L}^{-1}[G](t)$$

where a and b are arbitrary scalars, and $F(s)$ and $G(s)$ are the Laplace transforms of some functions.

We can write

$$y(t) = \mathcal{L}^{-1}\left[\frac{s-2}{s^2 - s - 6}\right](t) = \mathcal{L}^{-1}\left[\frac{s-2}{(s-3)(s+2)}\right](t).$$

Chapter 4: The Laplace Transform

For the moment we do not have a general method of finding the inverse Laplace transform of a given function. Nonetheless, we can use previous results in our specific example. To this end, we first decompose the rational fraction on the right-hand side into a sum of partial fractions:

$$\frac{s-2}{(s-3)(s+2)} = \frac{1/5}{s-3} + \frac{4/5}{s+2}.$$

We now use the linearity and the table of the last section to find the inverse Laplace transform of each of these simple fractions. That is

$$y(t) = \mathcal{L}^{-1}\left[\frac{s-2}{(s-3)(s+2)}\right](t)$$

$$= \mathcal{L}^{-1}\left[\frac{1/5}{s-3} + \frac{4/5}{s+2}\right](t)$$

$$= \tfrac{1}{5}\mathcal{L}^{-1}\left[\frac{1}{s-3}\right](t) + \tfrac{4}{5}\mathcal{L}^{-1}\left[\frac{1}{s+2}\right](t)$$

$$= \tfrac{1}{5}e^{3t} + \tfrac{4}{5}e^{-2t}.$$

Recall that the expression $s^2 - s - 6$ is the *characteristic polynomial* of the equation, and the functions e^{3t} and e^{-2t} are *fundamental solutions* of the equation.

It is sometimes not easy to solve problems of this sort. However, it is generally simpler to verify if the particular answer obtained is indeed a solution of the problem. We recommend checking your answer.

Example 4.10: We solve the initial value problem

$$\begin{cases} y''(t) - 2y'(t) + y(t) = 0, & t > 0, \\ y(0) = 1, \quad y'(0) = -2. \end{cases}$$

Solution: We apply the Laplace transform to both sides of the equation to obtain

$$0 = \mathcal{L}[y'' - 2y' + y]$$

$$= \mathcal{L}[y''] - 2\mathcal{L}[y'] + \mathcal{L}[y]$$

$$= \left[s^2\mathcal{L}[y] - sy(0) - y'(0)\right] - 2\left[s\mathcal{L}[y] - y(0)\right] + \mathcal{L}[y]$$

$$= (s^2 - 2s + 1)\mathcal{L}[y] - s + 4.$$

Thus

$$\mathcal{L}[y] = \frac{s-4}{s^2 - 2s + 1} = \frac{s-4}{(s-1)^2}.$$

Decompose the fraction on the right into a sum of partial fractions of the form

$$\mathcal{L}[y] = \frac{1}{s-1} - \frac{3}{(s-1)^2}.$$

Taking the inverse Laplace transforms of both sides, we obtain

$$y(t) = \mathcal{L}^{-1}\left[\frac{1}{s-1} - \frac{3}{(s-1)^2}\right](t)$$

$$= \mathcal{L}^{-1}\left[\frac{1}{s-1}\right](t) - 3\mathcal{L}^{-1}\left[\frac{1}{(s-1)^2}\right](t)$$

$$= e^t - 3te^t.$$

We used Formulae 2 and 10 from the table.

Example 4.11: Consider the initial value problem

$$\begin{cases} y''(t) + 2y'(t) + 5y(t) = 0, & t > 0, \\ y(0) = 1, \quad y'(0) = 0. \end{cases}$$

Solution: Again apply the Laplace transform to both sides of the equation:

$$0 = \mathcal{L}[y'' + 2y' + 5y]$$

$$= \mathcal{L}[y''] + 2\mathcal{L}[y'] + 5\mathcal{L}[y]$$

$$= \left[s^2\mathcal{L}[y] - sy(0) - y'(0)\right] + 2\left[s\mathcal{L}[y] - y(0)\right] + 5\mathcal{L}[y]$$

$$= (s^2 + 2s + 5)\mathcal{L}[y] - s - 2$$

and obtain

$$\mathcal{L}[y] = \frac{s+2}{s^2 + 2s + 5}.$$

In this case we present two different methods of solving for $y(t)$.

Method 1: Write $s^2 + 2s + 5 = (s+1)^2 + 4$ (complete the square). Then

$$\frac{s+2}{s^2 + 2s + 5} = \frac{s+1}{(s+1)^2 + 4} + \frac{1}{(s+1)^2 + 4}.$$

The reason for considering this particular representation is to allow us to use the formulae

$$\mathcal{L}\left[e^{at}\cos bt\right](s) = \frac{s-a}{(s-a)^2 + b^2} \quad \text{and} \quad \mathcal{L}\left[e^{at}\sin bt\right](s) = \frac{b}{(s-a)^2 + b^2}.$$

Chapter 4: The Laplace Transform

Thus
$$y(t) = \mathcal{L}^{-1}\left[\frac{s+2}{s^2+2s+5}\right](t)$$
$$= \mathcal{L}^{-1}\left[\frac{s+1}{(s+1)^2+4}\right](t) + \frac{1}{2}\mathcal{L}^{-1}\left[\frac{2}{(s+1)^2+4}\right](t)$$
$$= e^{-t}\cos 2t + \frac{1}{2}e^{-t}\sin 2t.$$

Method 2: We write $s^2 + 2s + 5 = (s+1-2i)(s+1+2i)$, and proceed with the usual decomposition into a sum of partial fractions, namely,
$$\frac{s+2}{s^2+5s+5} = \frac{1/2 - i/4}{s+1-2i} + \frac{1/2 + i/4}{s+1+2i}.$$

Therefore
$$y(t) = \mathcal{L}^{-1}\left[\frac{s+2}{s^2+2s+5}\right](t)$$
$$= (\tfrac{1}{2} - \tfrac{i}{4})\mathcal{L}^{-1}\left[\frac{1}{s+1-2i}\right](t) + (\tfrac{1}{2} + \tfrac{i}{4})\mathcal{L}^{-1}\left[\frac{1}{s+1+2i}\right]$$
$$= (\tfrac{1}{2} - \tfrac{i}{4})e^{-(1-2i)t} + (\tfrac{1}{2} + \tfrac{i}{4})e^{-(1+2i)t}.$$

In this step we used Formula 3 from the table. The only "problem" with the last expression is that it involves complex functions, while we may wish to have the solution in its real form. Working out this form we obtain

$$y(t) = (\tfrac{1}{2} - \tfrac{i}{4})e^{-t}[\cos 2t + i\sin 2t] + (\tfrac{1}{2} + \tfrac{i}{4})e^{-t}[\cos 2t - i\sin 2t]$$
$$= e^{-t}\cos 2t + \tfrac{1}{2}e^{-t}\sin 2t.$$

This is quite naturally the same answer as obtained by Method 1.

We have solved three representative examples of homogeneous second order constant coefficient linear ordinary differential equations. In the first example the roots of the characteristic polynomial were both real and distinct. In the second example the roots were real but equal. In the third example the roots were complex (but conjugate to each other). These three examples represent the prototypes of such differential equations. The methods we described in these examples can be used to solve every other initial value problem of the above type.

Let us now describe the method in its general form. Let

$$\begin{cases} ay''(t) + by'(t) + cy(t) = 0, & t > 0, \\ y(0) = y_0, \quad y'(0) = y_1, \end{cases}$$

be a homogeneous equation, where a, b, c, y_0, and y_1 are given real constants, and $a \neq 0$. We apply the Laplace transform to both sides of the equation to obtain

$$(as^2 + bs + c)\mathcal{L}[y](s) - (as+b)y_0 - ay_1 = 0.$$

Thus

$$\mathcal{L}[y](s) = y_0 \frac{as+b}{as^2+bs+c} + y_1 \frac{a}{as^2+bs+c}.$$

Set

$$F(s) = \frac{as+b}{as^2+bs+c}, \qquad G(s) = \frac{a}{as^2+bs+c}.$$

Then

$$y(t) = y_0 \mathcal{L}^{-1}[F](t) + y_1 \mathcal{L}^{-1}[G](t).$$

Note that $\mathcal{L}^{-1}[F](t)$ is the particular solution of the equation with initial conditions

$$\mathcal{L}^{-1}[F](0) = 1, \qquad \mathcal{L}^{-1}[F]'(0) = 0,$$

while $\mathcal{L}^{-1}[G](t)$ is the particular solution of the equation with initial conditions

$$\mathcal{L}^{-1}[G](0) = 0, \qquad \mathcal{L}^{-1}[G]'(0) = 1.$$

One way to compute $\mathcal{L}^{-1}[F](t)$ and $\mathcal{L}^{-1}[G](t)$ is as follows: If $as^2 + bs + c = a(s-\alpha)(s-\beta)$ where α and β are not equal, then we can write F (and also G) in the form

$$F(s) = \frac{A}{s-\alpha} + \frac{B}{s-\beta}$$

for some constants A and B. Thus

$$\mathcal{L}^{-1}[F](t) = Ae^{\alpha t} + Be^{\beta t}.$$

If $as^2 + bs + c = a(s-\alpha)^2$, then we can write F (and also G) in the form

$$F(s) = \frac{A}{s-\alpha} + \frac{B}{(s-\alpha)^2}.$$

Thus

$$\mathcal{L}^{-1}[F](t) = Ae^{\alpha t} + Bte^{\alpha t}.$$

If $as^2 + bs + c = a(s-(\alpha+i\beta))(s-(\alpha-i\beta))$ where α and β are real, and $\beta \neq 0$, then we can write F (and also G) in the form

$$F(s) = \frac{A\beta}{(s-\alpha)^2 + \beta^2} + \frac{B(s-\alpha)}{(s-\alpha)^2 + \beta^2}.$$

Thus

$$\mathcal{L}^{-1}[F](t) = Ae^{\alpha t}\sin\beta t + Be^{\alpha t}\cos\beta t.$$

Exercises

1. Compute the Laplace transform of each of the following functions:

   (a) $\dfrac{3s - 14}{s^2 - 4s + 18}$

   (b) $\dfrac{1}{s^2(s^2 + 1)}$

   (c) $\dfrac{1}{s^2 - 3s + 2}$

   (d) $\dfrac{s^2}{(s^2 + 4)^2}$

   (e) $\dfrac{s}{(s^2 + 4)^2}$

   (f) $\ln\left(1 + \dfrac{1}{s^2}\right)$

   (g) $\dfrac{1}{(s^2 + 4)^2}$

   (h) $\dfrac{s^3}{s^4 - 16}$

   (i) $\dfrac{1}{s^4 + 1}$

   (j) $\dfrac{2s^2 + s - 10}{(s - 4)(s^2 + 2s + 2)}$

2. For each of the following differential equations with associated initial conditions, find a particular solution on the interval $[0, \infty)$.

   (a) $y'' + 3y' - 4y = 0,$ $\quad y(0) = 3, \quad y'(0) = -2$
   (b) $y'' - y = 0,$ $\quad y(0) = 1, \quad y'(0) = -1$
   (c) $y'' + 4y' + 4y = 0,$ $\quad y(0) = 1, \quad y'(0) = 0$
   (d) $y'' + 4y = 0,$ $\quad y(0) = 2, \quad y'(0) = 1$
   (e) $y'' + 4y' + 7y = 0,$ $\quad y(0) = 1, \quad y'(0) = 1$
   (f) $ty''(t) + 2y'(t) + ty(t) = 0,$ $\quad y(0) = 1, \quad y'(0) = 0$

4. The Heaviside and Dirac-Delta Functions

In this section we consider two particular functions, their Laplace transforms, and various associated applications.

The Heaviside Function

For each real $c \geq 0$ we define

$$u_c(t) = \begin{cases} 0, & 0 \leq t < c, \\ 1, & c \leq t. \end{cases}$$

We call this function the *Heaviside function*. This is a particularly simple function and it is easy to calculate its Laplace transform.

$$\mathcal{L}[u_c](s) = \int_0^\infty e^{-st} u_c(t)\, dt = \int_c^\infty e^{-st} dt = \left.\dfrac{e^{-st}}{-s}\right|_c^\infty = \dfrac{e^{-cs}}{s}, \qquad s > 0.$$

For $c = 0$ we have $u_0(t) = 1$ and $\mathcal{L}[u_0](s) = \frac{1}{s}$. This formula is not valid for $c < 0$ (why?). We use this formula to more easily calculate the Laplace transforms of similar functions.

Example 4.12: Let $0 < c < d < \infty$. Set

$$f(t) = \begin{cases} 1, & c \leq t < d, \\ 0, & t \notin [c, d). \end{cases}$$

Let us calculate the Laplace transform of f. Since $f = u_c - u_d$ we have

$$\mathcal{L}[f](s) = \mathcal{L}[u_c](s) - \mathcal{L}[u_d](s) = \frac{e^{-cs} - e^{-ds}}{s}, \qquad s > 0.$$

Note that the values of f at the points c and d may be altered. A change of f at a finite number of points has absolutely no effect on the value of its Laplace transform.

Example 4.13: Let $f(t) = [t]$ be the function of the greatest integer value less than or equal to t. That is, if $n \in \mathbb{Z}_+$ and $n \leq t < n + 1$, then $f(t) = n$. We may calculate the Laplace transform of f as follows. As is easily seen

$$f = u_1 + u_2 + u_3 + \cdots.$$

Thus

$$\mathcal{L}[f](s) = \sum_{n=1}^{\infty} \frac{e^{-ns}}{s} = \frac{1}{s} \sum_{n=1}^{\infty} (e^{-s})^n = \frac{e^{-s}}{s(1 - e^{-s})}, \qquad s > 0.$$

(We have used the formula for the sum of a geometric series.)

The importance of the functions $u_c(t)$ in Laplace transform theory is not as illustrated by these examples. Rather it is a consequence of the next theorem.

Theorem 4.5: *Let f be a function for which $\mathcal{L}[f](s)$ exists for $s > a$. Let $c > 0$. Then*

$$\mathcal{L}[u_c(t) f(t - c)](s) = e^{-cs} \mathcal{L}[f](s), \qquad s > a.$$

Proof:

$$\mathcal{L}[u_c(t) f(t - c)](s) = \int_0^{\infty} e^{-st} u_c(t) f(t - c) \, dt = \int_c^{\infty} e^{-st} f(t - c) \, dt$$

$$= \int_0^{\infty} e^{-s(c+v)} f(v) \, dv = e^{-cs} \int_0^{\infty} e^{-sv} f(v) \, dv = e^{-cs} \mathcal{L}[f](s). \qquad \blacksquare$$

Chapter 4: The Laplace Transform

In Figures 4.1 and 4.2 we see the relationship between the graph of $f(t)$ and the graph of $u_c(t)f(t-c)$. The latter is simply a shift of the former by c units to the right, whereas $u_c(t)f(t-c)$ equals zero on the interval $[0,c]$.

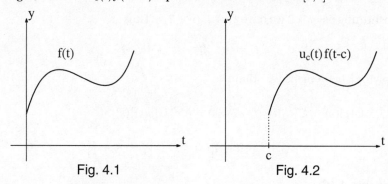

Fig. 4.1   Fig. 4.2

We will also use the converse of this theorem, i.e.,

(4.6) $$\mathcal{L}^{-1}\left[e^{-cs}\mathcal{L}[f](s)\right](t) = u_c(t)f(t-c).$$

Example 4.14: For $f(t) = t^2$ we calculate $\mathcal{L}[u_1(t)f(t-1)]$. We first remark that $u_1(t)f(t-1)$ is not the function $(t-1)^2$, nor does it equal t^2 for $t \geq 1$. To be precise

$$u_1(t)f(t-1) = \begin{cases} 0, & 0 \leq t < 1, \\ (t-1)^2, & t \geq 1. \end{cases}$$

From Theorem 4.5

$$\mathcal{L}[u_1(t)f(t-1)](s) = e^{-s}\mathcal{L}[f](s) = \frac{2e^{-s}}{s^3}, \qquad s > 0.$$

Example 4.15: Set

$$f(t) = \begin{cases} 0, & 0 \leq t < 2, \\ t^2, & 2 \leq t, \end{cases}$$

and let us calculate $\mathcal{L}[f]$. We may do this calculation using the definition of the Laplace transform. We have

$$\mathcal{L}[f](s) = \int_2^\infty e^{-st}t^2\, dt.$$

We can now integrate by parts twice and obtain the solution. However, it is easier and more edifying to rely on Theorem 4.5. To this end we search for the function g satisfying

$$f(t) = u_2(t)g(t-2).$$

For $t \geq 2$, the function g must satisfy

$$t^2 = f(t) = g(t-2).$$

Substituting $v = t - 2$ we have $g(v) = (v+2)^2$. Thus

$$g(t) = (t+2)^2 = t^2 + 4t + 4.$$

Applying Theorem 4.5 we obtain

$$\mathcal{L}[f](s) = \mathcal{L}[u_2(t)g(t-2)](s) = e^{-2s}\mathcal{L}[g(t)](s)$$

$$= e^{-2s}\mathcal{L}[t^2 + 4t + 4](s) = e^{-2s}\left[\tfrac{2}{s^3} + \tfrac{4}{s^2} + \tfrac{4}{s}\right], \qquad s > 0.$$

Example 4.16: Let

$$f(t) = \begin{cases} 5, & 0 \leq t < 1, \\ t+4, & 1 \leq t < 2, \\ 4t-2, & 2 \leq t. \end{cases}$$

We will calculate $\mathcal{L}[f]$. (Note that $\mathcal{L}[f]$ is not the sum of the Laplace transforms of the functions 5, $t+4$, and $4t-2$.) As in the previous example, it is possible to calculate $\mathcal{L}[f]$ directly from the definition by

$$\mathcal{L}[f](s) = \int_0^1 e^{-st} 5\, dt + \int_1^2 e^{-st}(t+4)\, dt + \int_2^\infty e^{-st}(4t-2)\, dt.$$

However, we will write f in the form

$$f(t) = g_0(t) + u_1(t)g_1(t-1) + u_2(t)g_2(t-2)$$

where g_0, g_1, g_2 are elementary functions, and then use Theorem 4.5. We must therefore identify g_0, g_1, g_2. On the interval $[0, 1)$ we have the equality

$$5 = f(t) = g_0(t).$$

This being so, we take $g_0(t) = 5$ for all $t \geq 0$. On the interval $[1, 2)$ we must have the equality

$$t + 4 = f(t) = 5 + g_1(t-1).$$

Thus $g_1(t-1) = t - 1$, from which we obtain $g_1(t) = t$. So let us set $g_1(t) = t$ on all of $[0, \infty)$. Finally, on the interval $[2, \infty)$ we have the equation

$$4t - 2 = f(t) = 5 + (t-1) + g_2(t-2).$$

Thus $g_2(t-2) = 3t - 6$, from which it follows that $g_2(t) = 3t$ for each $t \geq 0$. Finally,

$$\mathcal{L}[f](s) = \mathcal{L}[g_0](s) + e^{-s}\mathcal{L}[g_1](s) + e^{-2s}\mathcal{L}[g_2](s)$$

$$= \frac{5}{s} + \frac{e^{-s}}{s^2} + \frac{3e^{-2s}}{s^2}, \qquad s > 0.$$

One of the applications of Theorem 4.5 is in the solution of differential equations wherein there appear functions with a finite number of points of discontinuity.

Example 4.17: We will solve the equation

$$\begin{cases} y''(t) + y'(t) = h(t), & t > 0, \\ y(0) = 0, \quad y'(0) = 0, \end{cases}$$

where

$$h(t) = \begin{cases} 1, & \pi \leq t < 2\pi, \\ 0, & t \notin [\pi, 2\pi). \end{cases}$$

Since $h = u_\pi - u_{2\pi}$ we have

$$\mathcal{L}[y'' + y'](s) = \mathcal{L}[u_\pi - u_{2\pi}](s).$$

Simple calculations lead us to

$$(s^2 + s)\mathcal{L}[y](s) = \frac{e^{-\pi s} - e^{-2\pi s}}{s}.$$

Therefore

$$\mathcal{L}[y](s) = \frac{e^{-\pi s} - e^{-2\pi s}}{s^2(s+1)}$$

and

$$y(t) = \mathcal{L}^{-1}\left[\frac{e^{-\pi s}}{s^2(s+1)}\right](t) - \mathcal{L}^{-1}\left[\frac{e^{-2\pi s}}{s^2(s+1)}\right](t).$$

From (4.6)

$$\mathcal{L}^{-1}\left[\frac{e^{-\pi s}}{s^2(s+1)}\right](t) = u_\pi(t)g(t - \pi)$$

where

$$g(t) = \mathcal{L}^{-1}\left[\frac{1}{s^2(s+1)}\right](t).$$

In the same fashion

$$\mathcal{L}^{-1}\left[\frac{e^{-2\pi s}}{s^2(s+1)}\right](t) = u_{2\pi}(t)g(t - 2\pi).$$

It remains to calculate g. We decompose the fraction $\frac{1}{s^2(s+1)}$ into a sum of simple fractions of the form

$$\frac{1}{s^2(s+1)} = \frac{1}{s+1} - \frac{1}{s} + \frac{1}{s^2}.$$

We now apply Formulae 1, 2, and 6 from the table of formulae to obtain

$$g(t) = e^{-t} - 1 + t.$$

Thus

$$y(t) = u_\pi(t)\left[e^{-(t-\pi)} - 1 + (t-\pi)\right] - u_{2\pi}(t)\left[e^{-(t-2\pi)} - 1 + (t-2\pi)\right].$$

Remark: Note that $y(t) \equiv 0$ on the interval $[0, \pi]$. This is a consequence of the fact that the differential equation is homogeneous on that interval with homogeneous initial conditions. On the other hand, on the interval $[2\pi, \infty)$ we have $y(t) \neq 0$ despite the fact that the differential equation is also homogeneous on this interval (why?).

The Dirac-Delta Function

The function which we want to now briefly discuss is not in fact a function in the usual sense of the word. It is a function in a more complicated and general sense. However, we do not have here the mathematical tools necessary to properly and exactly define this more general type of a function. As such, we will present a somewhat imprecise definition of the Dirac-delta function. The Dirac-delta function, sometimes called the *impulse function*, at a point a is denoted by δ_a. In the special case where $a = 0$ we often write δ (in place of δ_0). We start with the "definition" of δ_a.

"Definition": *Let a be a given real number. The "function" δ_a satisfies*

$$\int_A f(t)\delta_a(t)\,dt = f(a)$$

for every function f which is continuous in a neighbourhood of a, and for each set A which contains a neighbourhood of a. This "function" is called the Dirac-delta function *at the point a.*

It is not hard to prove that no standard function can satisfy the above "definition". From this "definition" δ_a would have to vanish at each $t \neq a$, and have integral 1. This is, of course, impossible. (δ_a may be thought of as a

Chapter 4: The Laplace Transform

mathematical idealization of an instantaneous charge of mass one at the point a.) From this "definition" it does however follow that

$$\int_{-\infty}^{t} \delta_a(s)\, ds = \left\{ \begin{array}{ll} 0, & t < a, \\ 1, & t > a, \end{array} \right\} = u_a(t)$$

and therefore, in some sense, we can say that "$\delta_a(t) = u_a'(t)$".

We can easily calculate the Laplace transform of δ_a for each $a > 0$:

$$\mathcal{L}[\delta_a](s) = \int_0^\infty e^{-st} \delta_a(t)\, dt = e^{-as}.$$

We will not define the Laplace transform of δ_0. The function δ_a is important in signal theory and in the solution of associated differential equations. Let us present a simple example.

Example 4.18: Solve the initial value problem

$$\begin{cases} y''(t) + 2y'(t) + 2y(t) = \delta_\pi(t), & t > 0, \\ y(0) = 0, \quad y'(0) = 0. \end{cases}$$

Solution: Applying the Laplace transform to both sides of the equation we obtain

$$(s^2 + 2s + 2)\mathcal{L}[y] = e^{-\pi s}.$$

Thus

$$y(t) = \mathcal{L}^{-1}\left[\frac{e^{-\pi s}}{s^2 + 2s + 2}\right](t).$$

From Theorem 4.5 (see also (4.6))

$$y(t) = u_\pi(t) g(t - \pi)$$

where

$$g(t) = \mathcal{L}^{-1}\left[\frac{1}{s^2 + 2s + 2}\right](t) = \mathcal{L}^{-1}\left[\frac{1}{(s+1)^2 + 1}\right](t) = e^{-t} \sin t.$$

Thus

$$y(t) = u_\pi(t) e^{-(t-\pi)} \sin(t - \pi).$$

Exercises

1. Calculate the Laplace transform of each of the following functions:

   (a) $f(t) = \begin{cases} 1, & 0 \leq t \leq T \\ 0, & t > T \end{cases}$

   (b) $f(t) = \begin{cases} 0, & 0 \leq t \leq 1 \\ t^2 - t, & t > 1 \end{cases}$

   (c) $f(t) = \begin{cases} t^2, & 0 \leq t < 2 \\ 4t, & t \geq 2 \end{cases}$

   (d) $f(t) = \begin{cases} t, & 0 \leq t \leq 7 \\ 3 - t, & 7 < t < 8 \\ 1, & 8 \leq t \end{cases}$

2. Calculate the inverse Laplace transform of each of the following functions:

   (a) $\dfrac{e^{-s}(1 - e^{-s})}{s(s^2 + 1)}$

   (b) $\dfrac{e^{-16s}}{s(s^2 + 2s + 4)}$

3. Prove that $\mathcal{L}[\cos t \ln t \, \delta_\pi(t)] = -e^{-\pi s} \ln \pi$.

4. For each of the following differential equations with associated initial conditions, find the solution on the interval $[0, \infty)$.

   (a) $y''(t) + 4y'(t) + 7y(t) = u_1(t)$, $\quad y(0) = 1, \quad y'(0) = 0$

   (b) $y''(t) + 2y'(t) + 3y(t) = \delta_\pi(t)$, $\quad y(0) = 1, \quad y'(0) = 0$

   (c) $y''(t) - 2y'(t) + y(t) = (-1)^{[t]}$, $\quad y(0) = 0, \quad y'(0) = 0$

   (d) $y''(t) + 2y'(t) + 2y(t) = \delta_\pi(t)$, $\quad y(0) = 1, \quad y'(0) = 0$

   (e) $y''(t) + 4y(t) = \delta_\pi(t) - \delta_{2\pi}(t)$, $\quad y(0) = 0, \quad y'(0) = 0$

   (f) $y'''(t) - y(t) = \begin{cases} 1, & \pi \leq t \leq 2\pi, \\ 0, & \text{otherwise}, \end{cases}$ $\quad y(0) = y'(0) = 0, \quad y''(0) = 1$

5. Solve the finite differential-difference equation $y'(t) + y(t - 1) = t^2, t > 0$, where $y(t) = 0$ for every $t \leq 0$. (Hint: Use power series.)

5. Convolution

For the Fourier transform we proved that

$$\mathcal{F}[f * g](\omega) = 2\pi \mathcal{F}[f](\omega) \cdot \mathcal{F}[g](\omega).$$

The same formula is true for the Laplace transform without the 2π, i.e.,

$$\mathcal{L}[f * g](s) = \mathcal{L}[f](s) \cdot \mathcal{L}[g](s).$$

Chapter 4: The Laplace Transform

But first we need to understand what we mean by convolution here. In Chapter 3 we defined
$$(f * g)(t) = \int_{-\infty}^{\infty} f(t-y)g(y)\,dy.$$
In this chapter our functions are not defined on the interval $(-\infty, 0)$. To use the above definition we must extend their domain of definition to include this interval. We do this by setting $f(t) = 0$ for every $t < 0$. This being so, we may now write
$$(f * g)(t) = \int_{-\infty}^{\infty} f(t-y)g(y)\,dy.$$
Since $g(y) = 0$ for each $y < 0$, and $f(t-y) = 0$ for each $t < y$, we have

$$(f * g)(t) = \int_{-\infty}^{\infty} f(t-y)g(y)\,dy = \begin{cases} 0, & t < 0, \\ \int_0^t f(t-y)g(y)\,dy, & t \geq 0. \end{cases}$$

Hence for every f and g defined on the interval $[0, \infty)$, we define
$$(f * g)(t) = \int_0^t f(t-y)g(y)\,dy, \qquad t \geq 0,$$
for each t for which the integral exists. A simple substitution yields
$$(f * g)(t) = (g * f)(t).$$

Theorem 4.6: *If there exist constants a, K_1, and K_2 such that $|f(t)| \leq K_1 e^{at}$ and $|g(t)| \leq K_2 e^{at}$ for each $t \geq 0$, then*
$$|(f * g)(t)| \leq K_1 K_2 t e^{at}, \qquad t \geq 0,$$
and
$$\mathcal{L}[f * g](s) = \mathcal{L}[f](s) \cdot \mathcal{L}[g](s), \qquad s > a.$$

Proof:
$$|(f * g)(t)| = \left| \int_0^t f(t-y)g(y)\,dy \right| \leq \int_0^t |f(t-y)| \cdot |g(y)|\,dy$$
$$\leq \int_0^t K_1 e^{a(t-y)} K_2 e^{ay}\,dy \leq K_1 K_2 e^{at} \int_0^t dy$$
$$= K_1 K_2 t e^{at}.$$

Thus $e^{-st}(f * g)(t)$ is absolutely integrable for each $s > a$, and therefore the Laplace transform of $f * g$ exists for each $s > a$. Our assumptions imply that the functions $e^{-st}f(t)$ and $e^{-st}g(t)$ are absolutely integrable on $[0, \infty)$, for each $s > a$. Therefore, applying Fubini's Theorem (see Appendix B),

$$\mathcal{L}[f * g](s) = \int_0^\infty e^{-st} \left(\int_0^t f(t-y)g(y)\, dy \right) dt$$

$$= \int_0^\infty \left(\int_0^t e^{-s(t-y)} f(t-y) e^{-sy} g(y)\, dy \right) dt$$

$$= \int_0^\infty \left(\int_y^\infty e^{-s(t-y)} f(t-y) e^{-sy} g(y)\, dt \right) dy$$

$$= \int_0^\infty \left(\int_y^\infty e^{-s(t-y)} f(t-y)\, dt \right) e^{-sy} g(y)\, dy$$

$$= \int_0^\infty \left(\int_0^\infty e^{-su} f(u)\, du \right) e^{-sy} g(y)\, dy$$

$$= \mathcal{L}[f](s) \int_0^\infty e^{-sy} g(y)\, dy$$

$$= \mathcal{L}[f](s) \cdot \mathcal{L}[g](s). \qquad \blacksquare$$

One immediate and useful corollary of this theorem is that if $F = \mathcal{L}[f]$ and $G = \mathcal{L}[g]$ then

(4.7) $\qquad \mathcal{L}^{-1}\left[F(s)G(s)\right](t) = (f * g)(t) = \int_0^t f(t-y)g(y)\, dy.$

We now present a few simple applications of Theorem 4.6 and (4.7).

Example 4.19: Let f be a piecewise continuous function defined on $[0, \infty)$. Set

$$\varphi(t) = \int_0^t f(y)\, dy.$$

We wish to express $\mathcal{L}[\varphi]$ in terms of $\mathcal{L}[f]$. We do it by two different methods.

Method 1: It is easy to check that $\varphi(t) = (f * g)(t)$, where $g(t) \equiv 1$. Thus

$$\mathcal{L}[\varphi](s) = \mathcal{L}[f * g](s) = \mathcal{L}[f](s) \cdot \mathcal{L}[g](s) = \frac{\mathcal{L}[f](s)}{s}.$$

Method 2: Since $\varphi(0) = 0$ and $\varphi'(t) = f(t)$, we can use the derivative formula (Theorem 4.2) to obtain

$$\mathcal{L}[f](s) = \mathcal{L}[\varphi'](s) = s\mathcal{L}[\varphi] - \varphi(0) = s\mathcal{L}[\varphi](s)$$

and thus
$$\mathcal{L}[\varphi](s) = \frac{\mathcal{L}[f](s)}{s}.$$

Example 4.20: Let $F(s) = \frac{a}{s^2(s^2+a^2)}$, where $a > 0$. We calculate $\mathcal{L}^{-1}[F]$ by two different methods.

Method 1: We can write $F(s) = G(s)H(s)$, where $G(s) = \frac{1}{s^2}$ and $H(s) = \frac{a}{s^2+a^2}$. By Formula 6 from our table $\mathcal{L}^{-1}[G](t) = t$, and by Formula 4 $\mathcal{L}^{-1}[H](t) = \sin at$. Hence from Theorem 4.6 (or (4.7)) we have

$$\mathcal{L}^{-1}[F](t) = \int_0^t (t-y)\sin ay\, dy.$$

A simple integration (by parts) yields

$$\mathcal{L}^{-1}[F](t) = \frac{t}{a} - \frac{\sin at}{a^2}.$$

Method 2:
$$F(s) = \frac{a}{s^2(s^2+a^2)} = \frac{1/a}{s^2} - \frac{1/a}{s^2+a^2}$$

and thus

$$\mathcal{L}^{-1}[F](t) = \frac{1}{a}\mathcal{L}^{-1}\left[\frac{1}{s^2}\right](t) - \frac{1}{a^2}\mathcal{L}^{-1}\left[\frac{a}{s^2+a^2}\right](t) = \frac{t}{a} - \frac{\sin at}{a^2}.$$

Example 4.21: We want to find a function f which satisfies the integral equation

$$\int_0^t f(t-u)f(u)\, du = te^{-at}, \qquad t \geq 0.$$

Applying the Laplace transform to both sides of the equation we obtain

$$\mathcal{L}[f](s) \cdot \mathcal{L}[f](s) = \frac{1}{(s+a)^2}.$$

Thus $\mathcal{L}[f](s) = \pm\frac{1}{s+a}$, and therefore $f(t) = \pm e^{-at}$.

Example 4.22: We solve the integral equation

$$g(t) + \int_0^t g(u)e^{-(t-u)}\, du = 1, \qquad t \geq 0.$$

First we apply the Laplace transform to both sides of the equation to obtain

$$\mathcal{L}[g](s) + \mathcal{L}\left[g(t) * e^{-t}\right](s) = \mathcal{L}[1](s).$$

Thus

$$\mathcal{L}[g](s) + \mathcal{L}[g](s) \cdot \frac{1}{s+1} = \frac{1}{s}.$$

Solving for $\mathcal{L}[g]$ we have
$$\mathcal{L}[g](s) = \frac{s+1}{s(s+2)} = \frac{1/2}{s} + \frac{1/2}{s+2}.$$

Taking the inverse Laplace transform we obtain the solution to our integral equation,
$$g(t) = \frac{1}{2} + \frac{1}{2}e^{-2t}.$$

We end this section by using the convolution operation to present a general method for solving almost any second order linear constant coefficient ordinary differential equation on the interval $[0, \infty)$, with initial condition at $t = 0$. The general form of our problem is
$$\begin{cases} ay''(t) + by'(t) + cy(t) = h(t), & t > 0, \\ y(0) = y_0, \quad y'(0) = y_1, \end{cases}$$

where a, b, c, y_0, and y_1 are real constants, $a \neq 0$, and $h(t)$ is a given piecewise continuous function defined on $[0, \infty)$, whose Laplace transform exists for $s > a$, for some a. In Section 4.3 we found the general solution to the homogeneous equation. To solve the non-homogeneous equation it will thus suffice to solve
$$\begin{cases} ay''(t) + by'(t) + cy(t) = h(t), & t > 0, \\ y(0) = 0, \quad y'(0) = 0. \end{cases}$$

Apply the Laplace transform to both sides of the equation to obtain
$$(as^2 + bs + c)\mathcal{L}[y](s) = H(s)$$

where $H(s) = \mathcal{L}[h](s)$. Thus
$$\mathcal{L}[y](s) = \frac{H(s)}{as^2 + bs + c}.$$

Set $F(s) = \frac{1}{as^2+bs+c}$. By the methods of Section 4.3 we can find
$$f(t) = \mathcal{L}^{-1}[F](t).$$

The solution to our problem is therefore given by
$$y(t) = \mathcal{L}^{-1}[H(s)F(s)](t) = \int_0^t f(t-y)h(y)\,dy.$$

It is clear that we need not, in fact, compute $H(s)$.

Chapter 4: The Laplace Transform

Exercises

1. Calculate the Laplace transform of

$$f(t) = \int_0^t (u^2 - u + e^{-u})\, du.$$

2. Solve the following integral equations:

(a) $f(t) + 2\int_0^t f(u)\cos(t-u)\, du = 9e^{2t}$

(b) $\int_0^t f(u)\, du - f'(t) = \begin{cases} 0, & 0 \le t \le a, \\ 1, & a \le t, \end{cases} \quad a > 0$

(c) $f(t) + \int_0^t (t-u)f(u)\, du = \sin 2t$

(d) $f''(t) = \int_0^t u f(t-u)\, du, \quad f(0) = -1,\ f'(0) = 1$

(e) $f(t) + \int_0^t f(u)e^{-(t-u)}\, du = 1$

(f) $\int_0^t f'(u) f(t-u)\, du = 3te^{3t} - e^{3t} + 1, \quad f(0) = 0,\ f'(0) > 0$

(g) $3f'(t) - 10f(t) + 3\int_0^t f(u)\, du = 10\sin t - 5, \quad f(0) = 2$

(h) $\int_0^t f(u)f(t-u)\, du = 2f(t) + t - 2.$ (Is the solution unique?)

(i) $f'(t) + \int_0^t f(u)\, du = \sin t, \quad f(0) = 1$

(j) $\int_0^t f''(t) f(t-u)\, du = te^{at}, \quad f(0) = \frac{1}{a},\ f'(0) = 1$

(k) $f(t) = at + \int_0^t f(u)\sin(t-u)\, du$

(l) $f'(t) + 5\int_0^t f(u)\cos 2(t-u)\, du = 10, \quad f(0) = 2$

(j) $\int_0^t f'(u) f(t-u)\, du = 24t^3, \quad f(0) = 0$

3. For each of the following differential equations with associated initial conditions, find a solution on the interval $[0, \infty)$.

(a) $y''(t) + y(t) = g(t), \quad y(0) = 0,\ y'(0) = 0,$

where
$$g(t) = \begin{cases} t, & 0 \le t < 1, \\ 1, & 1 \le t. \end{cases}$$

(b) $y''(t) + 2y'(t) - 3y(t) = f(t)$, $y(0) = 1$, $y'(0) = 0$,

where
$$f(t) = \begin{cases} 0, & 0 \le t \le 2\pi, \\ \sin t, & t > 2\pi. \end{cases}$$

(c) $y''(t) + 2y'(t) + 3y(t) = \delta_\pi(t) + \sin t$, $y(0) = 0$, $y'(0) = 1$.

(d) $y''(t) + y(t) = \delta_\pi(t) \cos t$, $y(0) = 0$, $y'(0) = 1$.

(e) $y'''(t) - y''(t) + 4y'(t) - 4y(t) = 68e^t \sin 2t$,
$y(0) = 1$, $y'(0) = -19$, $y''(0) = -37$.

(f) $y''(t) + 9y(t) = f_c(t)$, $y(0) = a$, $y'(0) = b$,
where for $c > 0$,
$$f_c(t) = \begin{cases} 0, & 0 \le t \le c, \\ t - c, & c < t. \end{cases}$$

4. Let $y_1(t)$ be the solution of
$$y''(t) - 4y'(t) + 4y(t) = f_1(t), \qquad y(0) = 1, \quad y'(0) = 1, \quad t \ge 0,$$
and let $y_2(t)$ be the solution of
$$y''(t) - 4y'(t) + 4y(t) = f_2(t), \qquad y(0) = 1, \quad y'(0) = 1, \quad t \ge 0.$$
Suppose that $f_1(t) \le f_2(t)$, for each $t \ge 0$, and that f_1 and f_2 are continuous and bounded on the interval $[0, \infty)$. Prove that $y_1(t) \le y_2(t)$ for each $t \ge 0$.

6. More Examples and Applications

In this section we present some additional examples and applications of the Laplace transform. This section is independent of the subsequent sections and may be skipped.

Example 4.23: We calculate the Laplace transform of $f(t) = t^p$, $t \ge 0$, $p > -1$. It is easy to see that for every $p > -1$, $\mathcal{L}[f](s)$ is defined for each $s > 0$. Substitute $u = st$ ($s > 0$) to obtain
$$\mathcal{L}[f](s) = \int_0^\infty e^{-st} t^p dt = \int_0^\infty e^{-u} \left(\frac{u}{s}\right)^p \frac{du}{s} = \frac{1}{s^{p+1}} \int_0^\infty e^{-u} u^p du.$$

Chapter 4: The Laplace Transform

The value $\int_0^\infty e^{-u}u^p du$ is independent of s. It is a function of p and is called the *gamma function*. The standard definition of the gamma function is

$$\Gamma(p) = \int_0^\infty e^{-u}u^{p-1}du, \quad p > 0.$$

Thus we may write

$$\mathcal{L}[t^p](s) = \frac{\Gamma(p+1)}{s^{p+1}}, \quad s > 0, \quad p > -1.$$

The only problem is that we still do not know the value of $\Gamma(p+1)$ for all $p > -1$. However, since this result must conform with the formula obtained in Example 4.6, namely

$$\mathcal{L}[t^n](s) = \frac{n!}{s^{n+1}}, \quad s > 0, \quad n \in \mathbb{Z}_+,$$

we conclude that $\Gamma(n+1) = n!$ for each $n \in \mathbb{Z}_+$. This fact may be verified using the definition of the function Γ and integration by parts. It also follows from properties of the gamma function which we now present.

Properties of the Gamma Function

1. For every $p > 0$, $\Gamma(p+1) = p\Gamma(p)$.

Proof:

$$\Gamma(p+1) = \int_0^\infty e^{-u}u^p du$$

$$= -e^{-u}u^p \Big|_0^\infty - \int_0^\infty (-e^{-u})(pu^{p-1})du$$

$$= p\int_0^\infty e^{-u}u^{p-1}du$$

$$= p\Gamma(p). \qquad \blacksquare$$

2. $\Gamma(1) = 1$.

Proof:

$$\Gamma(1) = \int_0^\infty e^{-u}du = -e^{-u}\Big|_0^\infty = 1. \qquad \blacksquare$$

3. $\Gamma(\tfrac{1}{2}) = \sqrt{\pi}$.

Proof:

$$\Gamma(\tfrac{1}{2}) = \int_0^\infty e^{-u}u^{-\frac{1}{2}}du.$$

Make the substitution $v = u^{\frac{1}{2}}$. Then $dv = \tfrac{1}{2}u^{-\frac{1}{2}}du$, and

$$\Gamma(\tfrac{1}{2}) = 2\int_0^\infty e^{-v^2}dv = \int_{-\infty}^\infty e^{-v^2}dv.$$

In Chapter 3 (see Exercise 2 of Section 2) it is shown that this last integral equals $\sqrt{\pi}$. Thus $\Gamma(\frac{1}{2}) = \sqrt{\pi}$. ∎

From Properties 1 and 3 it follows that

$$\Gamma\left(\frac{2n+1}{2}\right) = \frac{(2n-1)(2n-3)\cdots 3 \cdot 1}{2^n}\sqrt{\pi}, \qquad n \in \mathbb{N},$$

and therefore

$$\mathcal{L}\left[\frac{1}{\sqrt{t}}\right](s) = \frac{\sqrt{\pi}}{\sqrt{s}}, \qquad \mathcal{L}\left[\sqrt{t}\right](s) = \frac{\sqrt{\pi}}{2s^{\frac{3}{2}}}, \qquad \ldots.$$

Example 4.24: The *beta function* is defined by

$$\mathcal{B}(p,q) = \int_0^1 x^{p-1}(1-x)^{q-1}dx, \qquad p,q > 0.$$

We use the Laplace transform to compute $\mathcal{B}(p,q)$ for every $p, q > 0$. Set

$$g(t) = \int_0^t x^{p-1}(t-x)^{q-1}dx.$$

Clearly $g(1) = \mathcal{B}(p,q)$ and $g(t)$ is the convolution of $f(t) = t^{p-1}$ and $h(t) = t^{q-1}$. Therefore

$$\mathcal{L}[g](s) = \mathcal{L}[f](s) \cdot \mathcal{L}[h](s) = \frac{\Gamma(p)}{s^p} \cdot \frac{\Gamma(q)}{s^q} = \frac{\Gamma(p)\Gamma(q)}{s^{p+q}}$$

and thus

$$g(t) = \mathcal{L}^{-1}\left[\frac{\Gamma(p)\Gamma(q)}{s^{p+q}}\right](t) = \frac{\Gamma(p)\Gamma(q)}{\Gamma(p+q)} \cdot \mathcal{L}^{-1}\left[\frac{\Gamma(p+q)}{s^{p+q}}\right](t)$$

$$= \frac{\Gamma(p)\Gamma(q)}{\Gamma(p+q)}t^{p+q-1}.$$

Hence

$$\int_0^t x^{p-1}(t-x)^{q-1}dx = \frac{\Gamma(p)\Gamma(q)}{\Gamma(p+q)}t^{p+q-1}$$

and in particular

$$\mathcal{B}(p,q) = \int_0^1 x^{p-1}(1-x)^{q-1}dx = \frac{\Gamma(p)\Gamma(q)}{\Gamma(p+q)}.$$

Example 4.25: In the proof of Theorem 3.3 we computed the integral $\int_0^\infty \frac{\sin x}{x}dx$. We now present another method of computing this integral. For each $t > 0$ define

$$f(t) = \int_0^\infty \frac{\sin tx}{x}dx.$$

Chapter 4: The Laplace Transform 171

Clearly our integral equals $f(1)$. (If we make the substitution $u = tx$ we see that f is a constant function. This fact will also follow from our calculations.)

$$\mathcal{L}[f](s) = \int_0^\infty e^{-st} \left(\int_0^\infty \frac{\sin tx}{x} dx \right) dt.$$

We change the order of integration (it is permitted in this case) to obtain

$$\mathcal{L}[f](s) = \int_0^\infty \left(\int_0^\infty e^{-st} \sin tx \, dt \right) \frac{dx}{x}$$

$$= \int_0^\infty \frac{x}{s^2 + x^2} \cdot \frac{1}{x} dx$$

$$= \int_0^\infty \frac{1}{s^2 + x^2} dx.$$

By the substitution $su = x$ we get

$$\mathcal{L}[f](s) = \frac{1}{s} \int_0^\infty \frac{1}{1+u^2} du = \frac{1}{s} \arctan u \Big|_0^\infty = \frac{\pi}{2s}.$$

Hence

$$f(t) = \mathcal{L}^{-1}\left[\frac{\pi}{2s}\right] = \frac{\pi}{2}$$

and in particular $f(1) = \frac{\pi}{2}$.

Example 4.26: The integrals $\int_0^\infty \cos u^2 \, du$ and $\int_0^\infty \sin u^2 \, du$ are called *Fresnel integrals*. We will prove that $\int_0^\infty \cos u^2 \, du = \frac{\sqrt{\pi}}{2\sqrt{2}}$. By the same method it may be shown that $\int_0^\infty \sin u^2 \, du = \frac{\sqrt{\pi}}{2\sqrt{2}}$. Set

$$f(t) = \int_0^\infty \cos tu^2 \, du.$$

Then for $s > 0$

$$\mathcal{L}[f](s) = \int_0^\infty e^{-st} \left(\int_0^\infty \cos tu^2 \, du \right) dt$$

$$= \int_0^\infty \left(\int_0^\infty e^{-st} \cos tu^2 \, dt \right) du$$

$$= \int_0^\infty \frac{s}{s^2 + u^4} du.$$

It is possible to justify the change in the order of integration, and the last line then follows from the fact that

$$\mathcal{L}\left[\cos tu^2\right](s) = \frac{s}{s^2 + (u^2)^2}.$$

Substituting $x = \frac{u}{\sqrt{s}}$, $s > 0$, we obtain

$$\mathcal{L}[f](s) = \int_0^\infty \frac{s}{s^2 + s^2 x^4} \sqrt{s}\, dx = \frac{1}{\sqrt{s}} \int_0^\infty \frac{1}{1+x^4}\, dx$$

$$= \frac{1}{2\sqrt{s}} \int_{-\infty}^\infty \frac{1}{1+x^4}\, dx.$$

Using partial fractions, or by the Residue Theorem (Theorem A.3), we can obtain

$$\int_{-\infty}^\infty \frac{1}{1+x^4}\, dx = \frac{\pi}{\sqrt{2}}.$$

Hence $\mathcal{L}[f](s) = \frac{\pi}{2\sqrt{2}\sqrt{s}}$. We already saw that $\mathcal{L}\left[\frac{1}{\sqrt{t}}\right](s) = \frac{\sqrt{\pi}}{\sqrt{s}}$. Thus

$$f(t) = \frac{\sqrt{\pi}}{2\sqrt{2}} \cdot \frac{1}{\sqrt{t}}$$

from which we immediately obtain

$$\int_0^\infty \cos u^2 \, du = f(1) = \frac{\sqrt{\pi}}{2\sqrt{2}}.$$

Exercises

1. Solve each of the following integral equations.

   (a) $\int_0^t \frac{f(u)}{\sqrt{t-u}}\, du = 1 + t + t^2$

   (b) $\int_0^t \frac{f(u)}{\sqrt[3]{t-u}}\, du = t(1+t)$

2. Prove, for every $p, q > 0$, $2\int_0^{\frac{\pi}{2}} \sin^{2p-1}\theta \cos^{2q-1}\theta \, d\theta = \mathcal{B}(p,q)$.

3. Evaluate the integrals

   (a) $\int_0^\infty x^2 e^{-2x^2}\, dx$,   (b) $\int_0^\infty \sqrt[4]{x} e^{-\sqrt{x}}\, dx$.

4. Prove, for every $p > 0$, $\int_0^1 \left(\ln\frac{1}{x}\right)^{p-1} dx = \Gamma(p)$.

5. Prove, for every $p, q > -1$, $\int_0^1 x^p \left(\ln\frac{1}{x}\right)^q dx = \frac{\Gamma(q+1)}{(p+1)^{q+1}}$.

6. Evaluate the integrals

   (a) $\int_0^2 (4-x^2)^{\frac{3}{2}}\, dx$,   (b) $\int_0^\infty \frac{1-\cos x}{x^2}\, dx$.

7. The *Riemann zeta function* is defined for each $p > 1$ by

$$\zeta(p) = \frac{1}{1^p} + \frac{1}{2^p} + \frac{1}{3^p} + \cdots.$$

Prove that

$$\zeta(p) = \frac{1}{\Gamma(p)} \int_0^\infty \frac{t^{p-1}}{e^t - 1} \, dt.$$

7. The Inverse Transform Formula

In Section 3.4 we presented a formula for the inverse Fourier transform. In this section we are interested in obtaining a similar formula for the inverse Laplace transform. To this end, we will have to generalize the definition of the Laplace transform so that $\mathcal{L}[f](\sigma)$ will also be defined for complex values $\sigma \in \mathbb{C}$. In a sense we are repeating what was done at the end of Chapter 3.

Let $\sigma = s + iv$ be an arbitrary complex number, $s, v \in \mathbb{R}$. Let f be a complex-valued piecewise continuous function defined on the interval $[0, \infty)$. We set

$$F(\sigma) = \int_0^\infty e^{-\sigma t} f(t) \, dt$$

for each σ at which the integral converges. If $\sigma = s$ then $F(\sigma) = F(s) = \mathcal{L}[f](s)$.

Theorem 4.7: *If $e^{-at} f(t)$ is piecewise continuous and absolutely integrable on the interval $[0, \infty)$, then $F(\sigma)$ is analytic on the complex half-plane $\text{Re}(\sigma) > a$.*

The proof of this result totally parallels the proof of Theorem 3.11, where we can regard our function f as being defined on all of $\mathbb{R}$, and identically zero on $(-\infty, 0)$.

We recall that an analytic function is completely determined by its values on a relatively small set of points. For example, the function is completely determined by its values on any interval inside its domain of analyticity. Hence if $F(s) = \mathcal{L}[f](s)$, for each $s > a$, then $F(\sigma)$ is completely determined by $\mathcal{L}[f](s)$. This fact is rather important since it enables us to obtain $F(\sigma)$ immediately from $\mathcal{L}[f](s)$ without further calculation. For example, if $f(t) = 1$ then we know that $\mathcal{L}[f](s) = \frac{1}{s}$, for every real $s > 0$. Hence it must be that $F(\sigma) = \frac{1}{\sigma}$, for every $\sigma \in \mathbb{C}$, $\text{Re}(\sigma) > 0$. Similarly, if we know that $F(s) = \frac{e^{-ds}}{s^2 + bs + c}$ for every real $s > a$, then it must be that $F(\sigma) = \frac{e^{-d\sigma}}{\sigma^2 + b\sigma + c}$ for every $\sigma \in \mathbb{C}$, $\text{Re}(\sigma) > a$.

Theorem 4.8: **(The Inverse Transform Formula)** *Assume that $e^{-at} f(t)$ is absolutely integrable on the interval $[0, \infty)$. Assume also that $f(t)$ is continuous* and its one-sided derivatives exist at each point t. Let

$$F(\sigma) = \int_0^\infty e^{-\sigma t} f(t)\, dt, \qquad \sigma \in \mathbb{C}, \quad \text{Re}(\sigma) > a.$$

Then

$$f(t) = \lim_{M \to \infty} \frac{1}{2\pi i} \int_{s-iM}^{s+iM} e^{\sigma t} F(\sigma)\, d\sigma, \qquad t > 0$$

for each $s > a$.

Remark: In the latter integral, the integration is along any straight line of the form

$$L_s = \{ \sigma = s + iv \in \mathbb{C} \mid -\infty < v < \infty \}$$

for which $s > a$. That is, any straight line parallel to the imaginary axis of the complex plane, contained completely inside the domain $\text{Re}(\sigma) > a$ (on which F is analytic). Note that the value of this integral is independent of the choice of $s > a$. This formula is sometimes called the *Fourier-Mellin inversion formula*.

We note that in order to obtain f from $\mathcal{L}[f]$ we must perform two operations.

(a) First extend $\mathcal{L}[f](s)$ to $F(\sigma)$.

(b) Then calculate the integral of $\frac{1}{2\pi i} e^{\sigma t} F(\sigma)$ along one of the appropriate lines L_s.

Proof of Theorem 4.8: We will use the Fourier transform and its inverse. Let $s > a$ be fixed throughout the proof. Define the function

$$G(v) = F(s + iv), \qquad v \in \mathbb{R},$$

i.e.,

$$G(v) = \int_0^\infty e^{-(s+iv)t} f(t)\, dt = \int_0^\infty e^{-ivt} \left[e^{-st} f(t) \right] dt$$

$$= \frac{1}{2\pi} \int_{-\infty}^\infty e^{-ivt} g(t)\, dt$$

where

$$g(t) = \begin{cases} 2\pi e^{-st} f(t), & t \geq 0, \\ 0, & t < 0. \end{cases}$$

Clearly g is piecewise continuous and absolutely integrable. Hence

$$G(v) = \mathcal{F}[g](v).$$

Chapter 4: The Laplace Transform

Likewise the one-sided derivatives of g exist at each point $t \neq 0$. Therefore by the Inverse Fourier Transform Theorem (Theorem 3.3)

$$g(t) = \text{P.V.} \int_{-\infty}^{\infty} e^{ivt} \mathcal{F}[g](v)\, dv = \text{P.V.} \int_{-\infty}^{\infty} e^{ivt} G(v)\, dv, \qquad t \neq 0.$$

For every $t > 0$ we obtain

$$2\pi e^{-st} f(t) = \text{P.V.} \int_{-\infty}^{\infty} e^{ivt} F(s + iv)\, dv$$

and thus

$$f(t) = \text{P.V.} \frac{1}{2\pi} \int_{-\infty}^{\infty} e^{(s+iv)t} F(s + iv)\, dv.$$

The variable of integration is v, and integration is along $\mathbb{R}$. We now make the substitution $\sigma = s + iv$. Since $dv = \frac{d\sigma}{i}$ we obtain our formula

$$f(t) = \lim_{M \to \infty} \frac{1}{2\pi i} \int_{s-iM}^{s+iM} e^{\sigma t} F(\sigma)\, d\sigma. \qquad \blacksquare$$

At the point $t = 0$ the right side will converge to $f(0+)/2$.

8. Applications of the Inverse Transform

In contrast to the situation with the inverse Fourier transform, the use of the inverse Laplace transform formula is less convenient. The calculation of the integral $\int_{s-i\infty}^{s+i\infty} e^{\sigma t} F(\sigma)\, d\sigma$ can be rather complicated. In this section we present one of the simple and natural methods of computing integrals of this form. The method is based on the Residue Theorem (Theorem A.3) and is very similar to the method of Section 6 of Chapter 3.

Suppose that $F(\sigma)$ is analytic on the domain $\text{Re}(\sigma) > a$, and $s > a$. We wish to compute

$$f(t) = \lim_{M \to \infty} \frac{1}{2\pi i} \int_{s-iM}^{s+iM} e^{\sigma t} F(\sigma)\, d\sigma, \qquad t > 0.$$

As mentioned earlier, there is no general method of doing this. We can, however, succeed in evaluating this integral under certain conditions on F. Suppose that $F(\sigma)$ is analytic on the entire complex plane, except perhaps at the finite number of points $\sigma_1, \sigma_2, \ldots, \sigma_n$, satisfying

$$\text{Re}(\sigma_j) < a, \qquad j = 1, 2, \ldots, n.$$

Let $s > a$, and let $R > 0$ be a real number sufficiently large that the right half-circle γ_R, with centre $(s, 0)$ and radius R, encloses all the points $\sigma_1, \sigma_2, \ldots, \sigma_n$.

Divide γ_R into the two segments

$$I_R = \{\sigma \in \mathbb{C} \mid \sigma = s + iv, \quad -R < v < R\},$$

$$C_R = \{\sigma \in \mathbb{C} \mid |\sigma - s| = R, \quad \operatorname{Re}(\sigma) \leq s\}$$

(as in Figure 4.3). By the Residue Theorem (Theorem A.3)

$$\frac{1}{2\pi i} \oint_{\gamma_R} e^{\sigma t} F(\sigma) \, d\sigma = \sum_{j=1}^{n} \operatorname{Res}\{e^{\sigma t} F(\sigma); \sigma_j\}.$$

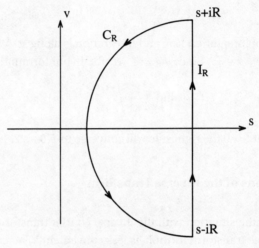

Fig. 4.3

The right side is independent of R, if R is sufficiently large. From $\gamma_R = C_R \cup I_R$ it follows that

$$\frac{1}{2\pi i} \oint_{\gamma_R} e^{\sigma t} F(\sigma) \, d\sigma = \frac{1}{2\pi i} \int_{I_R} e^{\sigma t} F(\sigma) \, d\sigma + \frac{1}{2\pi i} \int_{C_R} e^{\sigma t} F(\sigma) \, d\sigma$$

(where the integrals on the right have directions inherited from γ_R). Clearly

$$\lim_{M \to \infty} \frac{1}{2\pi i} \int_{s-iM}^{s+iM} e^{\sigma t} F(\sigma) \, d\sigma = \lim_{R \to \infty} \frac{1}{2\pi i} \int_{I_R} e^{\sigma t} F(\sigma) \, d\sigma.$$

Therefore if, by chance, we have

(4.8) $$\lim_{R \to \infty} \frac{1}{2\pi i} \int_{C_R} e^{\sigma t} F(\sigma) \, d\sigma = 0$$

then we obtain the formula

$$f(t) = \lim_{M \to \infty} \frac{1}{2\pi i} \int_{s-iM}^{s+iM} e^{\sigma t} F(\sigma) \, d\sigma = \sum_{j=1}^{n} \operatorname{Res}\{e^{\sigma t} F(\sigma); \sigma_j\}.$$

Chapter 4: The Laplace Transform

Unfortunately (4.8) does not hold for every F. In the next theorem we present a sufficient condition on F under which (4.8) holds.

Theorem 4.9: *Let F be an analytic function on the complex plane, except perhaps at a finite number of points. Let C_R be as above. If*

$$\lim_{R \to \infty} \max_{\sigma \in C_R} |F(\sigma)| = 0$$

then

$$\lim_{R \to \infty} \frac{1}{2\pi i} \int_{C_R} e^{\sigma t} F(\sigma) \, d\sigma = 0$$

holds for every $t > 0$.

This theorem is a reinterpretation of Proposition 3.6.

An immediate consequence of this theorem is the following. If F is analytic on $\mathbb{C}$, except perhaps at a finite number of points $\sigma_1, \sigma_2, \ldots, \sigma_n$, satisfying $\text{Re}(\sigma_j) < s$, $1 \leq j \leq n$, and if

$$\lim_{R \to \infty} \max_{\sigma \in C_R} |F(\sigma)| = 0,$$

then the inverse Laplace transform of $F(s)$ is

$$(4.9) \qquad f(t) = \lim_{M \to \infty} \frac{1}{2\pi i} \int_{s-iM}^{s+iM} e^{\sigma t} F(\sigma) \, d\sigma = \sum_{j=1}^{n} \text{Res}\{e^{\sigma t} F(\sigma); \sigma_j\}.$$

Example 4.27: We calculate the inverse Laplace transform of $F(s) = \frac{1}{s^2 - 3s + 2}$ by two different methods. In the first method we rewrite F using partial fractions as

$$F(s) = \frac{1}{s^2 - 3s + 2} = \frac{1}{s-2} - \frac{1}{s-1}.$$

Using known formulae we immediately obtain

$$f(t) = \mathcal{L}^{-1}\left[F(s)\right](t) = e^{2t} - e^{t}.$$

In the second method we compute $f(t)$ by using (4.9). From the remark after Theorem 4.7 it follows that $F(\sigma) = \frac{1}{\sigma^2 - 3\sigma + 2}$, where $\sigma \in \mathbb{C}$. The function $F(\sigma)$ has two simple poles $\sigma_1 = 1$ and $\sigma_2 = 2$ and therefore

$$f(t) = \lim_{M \to \infty} \frac{1}{2\pi i} \int_{s-iM}^{s+iM} e^{\sigma t} \frac{1}{\sigma^2 - 3\sigma + 2} \, d\sigma$$

where $s > 2$ is arbitrary. If we choose $s = 3$, then

$$C_R = \left\{ \sigma \in \mathbb{C} \,\Big|\, |\sigma - 3| = R, \quad \text{Re}(\sigma) \leq 3 \right\}.$$

We must check that $\lim_{R\to\infty} \max_{\sigma\in C_R} |F(\sigma)| = 0$ before we can use (4.9). Now

$$\max_{\sigma\in C_R} |F(\sigma)| = \max_{\sigma\in C_R} \left|\frac{1}{(\sigma-1)(\sigma-2)}\right|$$

If we let $R = |\sigma - 3|$ go to infinity then $|\sigma - 1|$ and $|\sigma - 2|$ will also converge to infinity, and therefore

$$\lim_{R\to\infty} \max_{\sigma\in C_R} \frac{1}{|(\sigma-1)(\sigma-2)|} = 0.$$

Thus

$$f(t) = \lim_{M\to\infty} \frac{1}{2\pi i} \int_{s-iM}^{s+iM} e^{\sigma t} \frac{1}{\sigma^2 - 3\sigma + 2} \, d\sigma$$

$$= \text{Res}\left\{\frac{e^{\sigma t}}{(\sigma-1)(\sigma-2)}; \sigma = 1\right\} + \text{Res}\left\{\frac{e^{\sigma t}}{(\sigma-1)(\sigma-2)}; \sigma = 2\right\}$$

$$= \left.\frac{e^{\sigma t}}{\sigma-2}\right|_{\sigma=1} + \left.\frac{e^{\sigma t}}{\sigma-1}\right|_{\sigma=2}$$

$$= \frac{e^t}{1-2} + \frac{e^{2t}}{2-1} = -e^t + e^{2t}.$$

We have, of course, obtained the exact same result as in the first method.

Example 4.28: For the Heaviside function $u_c(t)$, $c > 0$, we proved that

$$\mathcal{L}[u_c](s) = \frac{e^{-cs}}{s}, \quad s > 0.$$

If we set $F(s) = \frac{e^{-cs}}{s}$ then $F(\sigma) = \frac{e^{-c\sigma}}{\sigma}$. From Theorem 4.8 we obtain the formula

$$u_c(t) = \lim_{M\to\infty} \frac{1}{2\pi i} \int_{s-iM}^{s+iM} \frac{e^{\sigma t} e^{-c\sigma}}{\sigma} \, d\sigma, \quad 0 < t \neq c.$$

In contrast to the previous example, we cannot use (4.9) to obtain $u_c(t)$. This is because the function $\frac{e^{-c\sigma}}{\sigma}$ does not satisfy the condition of Theorem 4.9. It is easy to prove that

$$\lim_{R\to\infty} \max_{\sigma\in C_R} \left|\frac{e^{-c\sigma}}{\sigma}\right| = \infty.$$

If we set $\sigma = s - R$ then

$$\max_{\sigma\in C_R} \left|\frac{e^{-c\sigma}}{\sigma}\right| \geq \frac{e^{cR}e^{-cs}}{|s-R|}.$$

Since $c > 0$ the right-hand side tends to infinity as R tends to infinity. Thus, we cannot use (4.9) to find the inverse Laplace transform of $\frac{e^{-cs}}{s}$. Using (4.9) in this case leads to a wrong result. The function $\frac{e^{-c\sigma}}{\sigma}$ has a single simple pole at $\sigma = 0$, and therefore

$$\text{Res}\left\{\frac{e^{\sigma t}e^{-c\sigma}}{\sigma}; \sigma = 0\right\} = 1$$

for each value of t. This is, of course, not the function $u_c(t)$. We provide this example as a warning. Be careful in your use of (4.9)!

Example 4.29: We calculate the inverse Laplace transform of $F(s) = \frac{1}{s^3-1}$. Clearly $F(\sigma) = \frac{1}{\sigma^3-1}$ and the conditions of Theorem 4.9 are satisfied. Thus we may use (4.9). Set $f(t) = \mathcal{L}^{-1}\left[F(s)\right](t)$. Then

$$f(t) = \sum_{j=1}^{n} \text{Res}\left\{\frac{e^{\sigma t}}{\sigma^3 - 1}; \sigma_j\right\},$$

where $\sigma_1, \sigma_2, \ldots, \sigma_n$ are the poles of $\frac{1}{\sigma^3-1}$. In this case F has exactly three simple poles, $\sigma_1 = 1$, $\sigma_2 = \frac{-1+i\sqrt{3}}{2}$, and $\sigma_3 = \frac{-1-i\sqrt{3}}{2}$, and the corresponding residues are

$$\text{Res}\left\{\frac{e^{\sigma t}}{\sigma^3 - 1}; \sigma_j\right\} = \frac{e^{\sigma_j t}}{3\sigma_j^2}, \quad j = 1, 2, 3.$$

Thus

$$f(t) = \frac{e^t}{3 \cdot 1^2} + \frac{e^{\frac{-1+i\sqrt{3}}{2}t}}{3\left(\frac{-1+i\sqrt{3}}{2}\right)^2} + \frac{e^{\frac{-1-i\sqrt{3}}{2}t}}{3\left(\frac{-1-i\sqrt{3}}{2}\right)^2}.$$

By additional calculations we may arrive at the real form of f:

$$f(t) = \frac{e^t}{3} + \frac{(-1+i\sqrt{3})}{6}e^{\frac{-1+i\sqrt{3}}{2}t} + \frac{(-1-i\sqrt{3})}{6}e^{\frac{-1-i\sqrt{3}}{2}t}$$

$$= \frac{e^t}{3} - \frac{1}{3}e^{-\frac{t}{2}}\cos\frac{\sqrt{3}}{2}t - \frac{1}{\sqrt{3}}e^{-\frac{t}{2}}\sin\frac{\sqrt{3}}{2}t.$$

Another method of obtaining f is by the method of partial fractions.

Exercises

1. Use the inverse Laplace transform formula to calculate:

    (a) $\mathcal{L}^{-1}\left[\frac{\cosh a\sqrt{s}}{s \cosh \sqrt{s}}\right], 0 < a < 1$ (b) $\mathcal{L}^{-1}\left[\frac{1}{(1+s^2)^2}\right]$

    (c) $\mathcal{L}^{-1}\left[\frac{1}{s^3(1+s^2)}\right]$ (d) $\mathcal{L}^{-1}\left[\frac{s}{(1+s^2)^3}\right]$

(e) $\mathcal{L}^{-1}\left[\dfrac{1}{s\sqrt{1+s}}\right]$ (f) $\mathcal{L}^{-1}\left[\dfrac{\sqrt{s}}{s-1}\right]$

(g) $\mathcal{L}^{-1}\left[\dfrac{1}{s(e^s+1)}\right]$ (h) $\mathcal{L}^{-1}\left[\dfrac{1}{(s+2)^2(s^2+4)}\right]$

Review Exercises

1. Let $f(t) = \int_0^\infty \dfrac{u \sin ut}{1+u^2}\,du$.
   (a) Compute the Laplace transform of f.
   (b) Find f.

2. For each $c > 0$ let f_c be the $2c$-periodic function on the interval $[0, \infty)$ for which
$$f_c(t) = \begin{cases} t, & 0 \le t \le c, \\ 2c - t, & c \le t \le 2c. \end{cases}$$
   Calculate the Laplace transform of f_c.

3. Let f be a function such that $\dfrac{f(t)}{t}$ is piecewise continuous on $[0, \infty)$, and there exist constants K and a such that
$$|f(t)| \le Kte^{at}, \qquad t \ge 0.$$
   Prove that for each $s > a$,
$$\mathcal{L}\left[\dfrac{f(t)}{t}\right](s) = \int_s^\infty \mathcal{L}[f](u)\,du.$$

4. Prove that the Laplace transform of $\dfrac{1-e^{-t}}{t}$ is $\ln\left(1 + \dfrac{1}{s}\right)$.

5. Solve the integral equation
$$f(t) - \dfrac{1}{6}\int_0^t (t-y)^3 f(y)\,dy = t^2.$$

6. Let f and g be continuously differentiable and such that f, g, f', and g' are absolutely integrable on the interval $[0, \infty)$, and $f(0) = g(0) = 0$. Prove that
$$(f * g)' = f' * g = f * g'.$$

7. Compute the inverse Laplace transform of the functions
   (a) $\int_0^t (t-y)^2 \cos 2y\,dy$, (b) $\int_0^t \sin(t-y) \cos y\,dy$.

8. Solve the differential equation
$$y'' + 2y' + 2y = \sin at, \qquad y(0) = 0, \quad y'(0) = 0,$$
   where a is a given constant.

9. Find the inverse Laplace transform of $\frac{1}{(s+1)^2(s^2+4)}$.
10. Solve the following systems of equations:

    (a)
    $$\begin{cases} x'(t) + 2x(t) - 4y(t) = f(t), \\ y'(t) + x(t) - 2y(t) = 0, \\ x(0) = y(0) = 0, \end{cases}$$

    where $f(t) = \begin{cases} t, & 0 \le t \le 1, \\ 1, & 1 < t \le 2, \\ t-1, & 2 < t < \infty. \end{cases}$

    (b)
    $$\begin{cases} x'(t) + y(t) = 0, \\ x''(t) + y'(t) + y(t) = e^t, \\ x(0) = 1, \quad x'(0) = 0, \quad y(0) = 0. \end{cases}$$

11. Use the Laplace transform to solve the following systems of partial differential equations:

    (a)
    $$\begin{cases} u_x + 2xu_t = 2x, & 0 < x < \infty, \quad 0 < t < \infty, \\ u(x,0) = 1, & 0 \le x < \infty, \\ u(0,t) = 1, & 0 \le t < \infty. \end{cases}$$

    (b)
    $$\begin{cases} xu_x + u_t = xt, & 0 < x < \infty, \quad 0 < t < \infty, \\ u(x,0) = 0, & 0 \le x < \infty, \\ u(0,t) = 0, & 0 \le t < \infty. \end{cases}$$

Appendix A
The Residue Theorem and Related Results

Throughout this book we have assumed that the reader is conversant with the complex numbers $\mathbb{C}$, and Euler's formula

$$e^{iy} = \cos y + i \sin y, \quad y \in \mathbb{R},$$

and how to manipulate it.

This is not really enough. A proper understanding of the Fourier and Laplace transforms is not possible without a smattering of knowledge and comprehension of analytic function theory. In this appendix we review some of the concepts and results of analytic function theory (complex variables) used in this book. Most of the book can be read without this theory, but it is even better if read with some knowledge of this theory.

A *domain* $\mathcal{D}$ in $\mathbb{C}$ is any open set in $\mathbb{C}$. That is, it is a set such that if $z_0 \in \mathcal{D}$, then some disk centred at z_0, with positive radius, is contained in $\mathcal{D}$.

A function f defined on a domain $\mathcal{D}$ is said to be *analytic* thereon if its derivative (with respect to z) exists at every point in $\mathcal{D}$. For $z = x + iy$, $x, y \in \mathbb{R}$, let us write

$$f(z) = u(x, y) + iv(x, y)$$

for real-valued u and v. Then f is analytic on $\mathcal{D}$ if and only if the first partials of u and v exist, are continuous on $\mathcal{D}$, and satisfy the Cauchy-Riemann equations

$$u_x = v_y \quad \text{and} \quad u_y = -v_x.$$

If f is analytic on $\mathcal{D}$ then derivatives of all orders exist, and are continuous. Even more is true, namely, for each $z_0 \in \mathcal{D}$ there exists a disk of positive radius centred at z_0 such that on this disk

$$f(z) = \sum_{n=0}^{\infty} \frac{f^{(n)}(z_0)}{n!} (z - z_0)^n.$$

Appendix A: Residue Theorem and Related Results

This power series is called the *Taylor series* of f about the point z_0.

A continuous complex-valued function γ defined on the real closed interval $[a, b]$ is said to define a *curve* in $\mathbb{C}$. The curve is *simple* if $\gamma(s) \neq \gamma(t)$ for all $a \leq s < t < b$. (This just says that the curve does not come back on itself.) It is *closed* if $\gamma(a) = \gamma(b)$. Every simple closed curve divides $\mathbb{C}$ into three parts: the curve itself, the interior of the curve, and the exterior of the curve. Each simple closed curve is generally given an *orientation*. The orientation is anticlockwise (this is important as we will consider integrals along such curves).

A *simply connected domain* is a domain with no holes, or anything else missing. More formally, it is defined as a domain $\mathcal{D}$ in which for every simple closed curve γ in $\mathcal{D}$, the interior of γ is also in $\mathcal{D}$.

Two fundamental facts concerning integrals of analytic functions in simply connected domains are the following.

Theorem A.1: (**Cauchy's Theorem**) *If f is an analytic function on a simply connected domain $\mathcal{D}$, then for every simple closed curve γ in $\mathcal{D}$*

$$\oint_\gamma f(z)\,dz = 0.$$

Theorem A.2: (**Morera's Theorem**) *If f is continuous on a simply connected domain $\mathcal{D}$, and*

$$\oint_\gamma f(z)\,dz = 0$$

for every simple closed curve γ in $\mathcal{D}$, then f is analytic in $\mathcal{D}$.

Assume f is analytic in a simply connected domain $\mathcal{D}$ except that it is not analytic at one point z_0 in $\mathcal{D}$. This point is called a *singular point*. In some neighbourhood (disk of positive radius) about z_0, f can be represented by a *Laurent series*

$$f(z) = \sum_{n=0}^{\infty} a_n (z - z_0)^n + \sum_{n=1}^{\infty} b_n (z - z_0)^{-n}.$$

Let γ be a simple closed curve in $\mathcal{D}$ with z_0 in its interior. Then

$$\frac{1}{2\pi i} \oint_\gamma f(z)\,dz = b_1.$$

The value b_1 is called the *residue* of f at z_0. We denote it by $\text{Res}\{f; z_0\}$.

As a consequence we have

Theorem A.3: (**Residue Theorem**) *Let γ be a simple closed curve. Assume that f is analytic on γ, and interior to γ except at a finite number of points $z_1, \ldots, z_n$.*

Then

$$\frac{1}{2\pi i}\oint_\gamma f(z)\,dz = \sum_{k=1}^{n} \operatorname{Res}\{f; z_j\}.$$

How do we calculate $\operatorname{Res}\{f; z_k\}$ (other than by calculating the requisite integrals)?

Let z_0 be a singular point of f. Let

$$f(z) = \sum_{n=0}^{\infty} a_n(z-z_0)^n + \sum_{n=1}^{m} b_n(z-z_0)^{-n}.$$

denote the Laurent series of f about z_0, with m finite and $b_m \neq 0$. Then z_0 is said to be a *pole* of f of *order* m. Let

$$\phi(z) = (z-z_0)^m f(z).$$

Note that (if we define $\phi(z_0) = b_m$) ϕ may be considered as analytic at z_0. In addition, $\phi(z_0) \neq 0$. (If we multiply f by any other power of $(z-z_0)$ then one of the above two conditions does not hold.) Now

$$\phi(z) = \sum_{n=1}^{\infty} a_n(z-z_0)^{n+m} + \sum_{n=1}^{m} b_n(z-z_0)^{m-n}.$$

Since ϕ is analytic at z_0, it follows from the uniqueness of the Taylor series that

$$b_1 = \frac{\phi^{(m-1)}(z_0)}{(m-1)!}.$$

Summarizing we have: Let z_0 be a singular point of f. Assume m is a positive integer and

$$\phi(z) = (z-z_0)^m f(z)$$

may be defined at z_0 so that it is analytic there and satisfies $\phi(z_0) \neq 0$. Then f has a pole of order m at z_0, and

$$\operatorname{Res}\{f; z_0\} = \frac{\phi^{(m-1)}(z_0)}{(m-1)!}.$$

For $m = 1$, this reduces to

$$\operatorname{Res}\{f; z_0\} = \lim_{z \to z_0}(z-z_0)f(z).$$

In many cases singular points occur when f is of the form

$$f(z) = \frac{g(z)}{h(z)}$$

Appendix A: Residue Theorem and Related Results

where both g and h are analytic functions. The function f is analytic at all points z where g and h are analytic, and $h(z) \neq 0$. Assume $h(z_0) = 0$. Then f has a pole or a removable singularity at z_0. (A point of singularity is removable if the function may be redefined at that point such that the singularity does not then exist.) Assume g has a zero at z_0 of order r, i.e.,

$$g(z_0) = g'(z_0) = \cdots = g^{(r-1)}(z_0) = 0, \qquad g^{(r)}(z_0) \neq 0,$$

and h has a zero at z_0 of order s. If $r \geq s$, then z_0 is a removable singularity of f. If $s > r$, then f has a pole of order $s - r$ at z_0.

If $s = 1$ and $r = 0$, i.e., $g(z_0) \neq 0$, $h(z_0) = 0$, $h'(z_0) \neq 0$, then z_0 is a simple (order 1) pole and

$$\text{Res}\{f; z_0\} = \frac{g(z_0)}{h'(z_0)}.$$

Appendix B

Leibniz's Rule and Fubini's Theorem

In some of the proofs of the theorems of Chapters 3 and 4 we interchanged the order of two successive operations (two successive integrations or differentiation followed by integration). In this appendix we state these theorems precisely. The first theorem (Leibniz's rule) describes conditions under which we are allowed to interchange the order of differentiation followed by integration (and vice versa). The second theorem, generally called Fubini's theorem, concerns conditions under which we are permitted to interchange the order of two successive integrations.

Definition B.1: *Let f be a complex-valued function of two independent real variables x and y. We call f piecewise continuous if for every fixed x_0, the function $g(y) = f(x_0, y)$ is a piecewise continuous function (as an ordinary function of the one variable y), and for every fixed y_0, the function $h(x) = f(x, y_0)$ is piecewise continuous.*

Definition B.2: *Let $f : \mathbb{R}^2 \to \mathbb{C}$ be a piecewise continuous function. The integral $\int_{-\infty}^{\infty} f(x, y)\, dx$ is said to converge uniformly on the interval $[c, d]$ if:*

(a) *The integral $\int_{-\infty}^{\infty} f(x, y)\, dx$ converges for every y in the interval $[c, d]$.*

(b) *For every $\epsilon > 0$ there exists a real number M_ϵ (which depends on ϵ), such that for all $M \geq M_\epsilon$ and every $c \leq y \leq d$ we have*

$$\left| \int_{-\infty}^{\infty} f(x, y)\, dx - \int_{-M}^{M} f(x, y)\, dx \right| < \epsilon.$$

We now state a theorem on the interchange of differentiation and integration.

Theorem B.3: *(Leibniz's rule) Let $f : \mathbb{R}^2 \to \mathbb{C}$ be continuous and differentiable with respect to y. If $\int_{-\infty}^{\infty} f(x, y)\, dx$ converges for every y, and $\int_{-\infty}^{\infty} \frac{\partial f}{\partial y}(x, y)\, dx$*

converges uniformly on every interval $[c,d]$, then the function $\int_{-\infty}^{\infty} f(x,y)\,dx$ is differentiable on $\mathbb{R}$ and

$$\frac{d}{dy}\int_{-\infty}^{\infty} f(x,y)\,dx = \int_{-\infty}^{\infty} \frac{\partial f}{\partial y}(x,y)\,dx.$$

We now state a form of Fubini's Theorem for interchanging the order of two successive integrations.

Theorem B.4: (Fubini) *Let $f : \mathbb{R}^2 \to \mathbb{C}$ be piecewise continuous. Suppose that $\int_{-\infty}^{\infty} |f(x,y)|\,dx$ converges uniformly on every interval $[c,d]$, and that $\int_{-\infty}^{\infty} |f(x,y)|\,dy$ converges uniformly on every interval $[a,b]$. If one of the integrals*

$$\int_{-\infty}^{\infty}\int_{-\infty}^{\infty} |f(x,y)|\,dy\,dx, \qquad \int_{-\infty}^{\infty}\int_{-\infty}^{\infty} |f(x,y)|\,dx\,dy$$

converges, then the two integrals

$$\int_{-\infty}^{\infty}\int_{-\infty}^{\infty} f(x,y)\,dy\,dx, \qquad \int_{-\infty}^{\infty}\int_{-\infty}^{\infty} f(x,y)\,dx\,dy$$

converge and

$$\int_{-\infty}^{\infty}\int_{-\infty}^{\infty} f(x,y)\,dy\,dx = \int_{-\infty}^{\infty}\int_{-\infty}^{\infty} f(x,y)\,dx\,dy.$$

Index

Absolutely integrable, 94
Analytic, 182

Band-limited, 130
Basis, 7
Bessel's inequality, 25, 50
Beta function, 170

Cauchy's Theorem, 183
Cauchy-Schwarz inequality, 11
Closed (orthonormal system), 26
Complete (orthonormal system), 27
Complex Fourier series, 43
Convergence in norm, 26
Convolution, 116, 163
Cosine series, 41, 74
Curve, 183
    closed, 183
    orientation, 183
    simple, 183

Derivative formula, 105, 143
Dirac-delta function, 160
Dirichlet kernel, 49
Dirichlet's Theorem, 47
Distance, 10, 20
Domain (in the complex plane), 182
    simply connected, 183

Euclidean norm, 11
Even function, 40

Fourier coefficients, 35
Fourier series, 35
    complex, 43
    on $[a,b]$, 81
Fourier transform, 93

Fourier-Mellin inversion formula, 174
Fresnel integrals, 171
Fubini's Theorem, 187

Gamma function, 169
Generalized Fourier coefficients, 17
Generalized Parseval identity, 28, 67
Generalized Plancherel identity, 114
Generalized Pythagorean Theorem, 17
Gibbs phenomenom, 68
Gram-Schmidt process, 22

Heaviside function, 155

Inner product, 7
    standard, 8
Inner product space, 7
Inverse Fourier transform, 108, 109
Inverse Laplace transform, 148, 173

Jordan's Lemma, 122
Jump point, 33

Laplace transform, 140
Laurent series, 183
Lebesgue Dominated Convergence
    Theorem, 94
Leibniz's rule, 186
Linear combination, 6
Linear space, 5
    dimension, 7
    subspace, 6
Linearly dependent, 6
Linearly independent, 6
Low-pass filter, 130

Modulation formulae, 104
Morera's Theorem, 183

Index

Norm, 10

Odd function, 40
Orthogonal, 15
Orthogonal projection, 19
Orthogonal system, 15
Orthonormal system, 15

Parseval's identity, 25, 27, 63
Perpendicular, 15
Piecewise continuous, 3, 32, 94, 186
Plancherel identity, 113
Pointwise convergence, 46, 56
Pole, 184
Principal value, 111

Rational function, 124
Residue, 183
Residue Theorem, 183
Riemann-Lebesgue Lemma, 25, 50

Shannon Sampling Theorem, 132
Shift formula, 103
Signal, 130
Sine series, 41, 74
Singular point, 183

Taylor series, 183
Time-limited, 130
Triangle inequality, 10
Trigonometric formulae, 4
Trigonometric polynomial, 37

Uniform convergence, 56
Uniform norm, 11
Uniqueness of Fourier series, 67
Unit vector, 15

Vector space, 5